PERILLA

Medicinal and Aromatic Plants – Industrial Profiles

Individual volumes in this series provide both industry and academia with in-depth coverage of one major medicinal or aromatic plant of industrial importance.

Edited by Dr Roland Hardman

Volume 1
Valerian
edited by Peter J. Houghton

Volume 2
Perilla
edited by He-Ci Yu, Kenichi Kosuna and Megumi Haga

Other volumes in preparation

Artemisia, edited by C. Wright
Cannabis, edited by D. Brown
Capsicum, edited by P. Bosland and A. Levy
Cardamom, edited by P.N. Ravindran and K.J. Madusoodanan
Carum, edited by É. Németh
Chamomile, edited by R. Franke and H. Schilcher
Cinnamon and Cassia, edited by P.N. Ravindran and S. Ravindran
Claviceps, edited by V. Křen and L. Cvak
Colchicum, edited by V. Simánek
Curcuma, edited by B.A. Nagasampagi and A.P. Purohit
Eucalyptus, edited by J. Coppen
Evening Primrose, edited by P. Lapinskas
Feverfew, edited by M.I. Berry
Ginkgo, edited by T. van Beek
Ginseng, by W. Court
Illicium and Pimpinella, edited by M. Miró Jodral
Licorice, by L.E. Craker and L. Kapoor
Melaleuca, edited by I. Southwell
Neem, by H.S. Puri
Ocimum, edited by R. Hiltunen and Y. Holt
Piper Nigrum, edited by P.N. Ravindran
Plantago, edited by C. Andary and S. Nishibe
Poppy, edited by J. Bernáth
Saffron, edited by M. Negbi
Stevia, edited by A.D. Kinghorn
Tilia, edited by K.P. Svoboda and J. Collins
Trigonella, edited by G.A. Petropoulos
Urtica, by G. Kavalali

This book is part of a series. The publisher will accept continuation orders which may be cancelled at any time and which provide for automatic billing and shipping of each title in the series upon publication. Please write for details.

PERILLA
The Genus *Perilla*

Edited by

He-Ci Yu
Hankintatukku Natural Products Co., Helsinki, Finland

Kenichi Kosuna
Amino Up Chemical Co., Sapporo, Japan

and

Megumi Haga
Amino Up Chemical Co., Sapporo, Japan

harwood academic publishers
Australia • Canada • China • France • Germany • India • Japan
Luxembourg • Malaysia • The Netherlands • Russia • Singapore
Switzerland • Thailand • United Kingdom

Copyright © 1997 OPA (Overseas Publishers Association) Amsterdam B.V. Published in The Netherlands by Harwood Academic Publishers.

Amsteldijk 166
1st Floor
1079 LH Amsterdam
The Netherlands

British Library Cataloguing in Publication Data

A catalogue record for this book is available from the British Library.

CONTENTS

PREFACE TO THE SERIES

There is increasing interest in industry, academia and the health sciences in medicinal and aromatic plants. In passing from plant production to the eventual product used by the public, many sciences are involved. This series brings together information which is currently scattered through an ever increasing number of journals. Each volume gives an in-depth look at one plant genus, about which an area specialist has assembled information ranging from the production of the plant to market trends and quality control.

Many industries are involved such as forestry, agriculture, chemical, food, flavour, beverage, pharmaceutical, cosmetic and fragrance. The plant raw materials are roots, rhizomes, bulbs, leaves, stems, barks, wood, flowers, fruits and seeds. These yield gums, resins, essential (volatile) oils, fixed oils, waxes, juices, extracts and spices for medicinal and aromatic purposes. All these commodities are traded world-wide. A dealer's market report for an item may say "Drought in the country of origin has forced up prices".

Natural products do not mean safe products and account of this has to be taken by the above industries, which are subject to regulation. For example, a number of plants which are approved for use in medicine must not be used in cosmetic products.

The assessment of safe to use starts with the harvested plant material which has to comply with an official monograph. This may require absence of, or prescribed limits of, radioactive material, heavy metals, aflatoxin, pesticide residue, as well as the required level of active principle. This analytical control is costly and tends to exclude small batches of plant material. Large scale contracted mechanised cultivation with designated seed or plantlets is now preferable.

Today, plant selection is not only for the yield of active principle, but for the plant's ability to overcome disease, climatic stress and the hazards caused by mankind. Such methods as *in vitro* fertilisation, meristem cultures and somatic embryogenesis are used. The transfer of sections of DNA is giving rise to controversy in the case of some end-uses of the plant material.

Some suppliers of plant raw material are now able to certify that they are supplying organically-farmed medicinal plants, herbs and spices. The Economic Union directive (CVO/EU No 2092/91) details the specifications for the **obligatory** quality controls to be carried out at all stages of production and processing of organic products.

Fascinating plant folklore and ethnopharmacology leads to medicinal potential. Examples are the muscle relaxants based on the arrow poison, curare, from species of *Chondrodendron*, and the antimalarials derived from species of *Cinchona* and *Artemisia*. The methods of detection of pharmacological activity have become increasingly reliable and specific, frequently involving enzymes in bioassays and avoiding the use of laboratory animals. By using bioassay linked fractionation of crude plant juices or extracts, compounds can be specifically targeted which, for example, inhibit blood platelet aggregation, or have antitumour, or antiviral, or any other required activity. With the assistance of robotic devices, all the members of a genus may be readily screened. However, the plant material must be **fully** authenticated by a specialist.

The medicinal traditions of ancient civilisations such as those of China and India have a large armamentaria of plants in their pharmacopoeias which are used throughout South East Asia. A similar situation exists in Africa and South America. Thus, a very high percentage of the World's population relies on medicinal and aromatic plants for their medicine. Western medicine is also responding. Already in Germany all medical practitioners have to pass an examination in phytotherapy before being allowed to practise. It is noticeable that throughout Europe and the USA, medical, pharmacy and health related schools are increasingly offering training in phytotherapy.

Multinational pharmaceutical companies have become less enamoured of the single compound magic bullet cure. The high costs of such ventures and the endless competition from me too compounds from rival companies often discourage the attempt. Independent phytomedicine companies have been very strong in Germany. However, by the end of 1995, eleven (almost all) had been acquired by the multinational pharmaceutical firms, acknowledging the lay public's growing demand for phytomedicines in the Western World.

The business of dietary supplements in the Western World has expanded from the Health Store to the pharmacy. Alternative medicine includes plant based products. Appropriate measures to ensure the quality, safety and efficacy of these either already exist or are being answered by greater legislative control by such bodies as the Food and Drug Administration of the USA and the recently created European Agency for the Evaluation of Medicinal Products, based in London.

In the USA, the Dietary Supplement and Health Education Act of 1994 recognised the class of phytotherapeutic agents derived from medicinal and aromatic plants. Furthermore, under public pressure, the US Congress set up an Office of Alternative Medicine and this office in 1994 assisted the filing of several Investigational New Drug (IND) applications, required for clinical trials of some Chinese herbal preparations. The significance of these applications was that each Chinese preparation involved several plants and yet was handled as a **single** IND. A demonstration of the contribution to efficacy, of **each** ingredient of **each** plant, was not required. This was a major step forward towards more sensible regulations in regard to phytomedicines.

My thanks are due to the staff of Harwood Academic Publishers who have made this series possible and especially to the volume editors and their chapter contributors for the authoritative information.

Roland Hardman

PREFACE

Herbal plants have served as a valuable resource, and have provided most of the therapeutic entities needed for traditional medicine. In recent decades, increasing attention has been paid to herbal plants by modern medicine. This is because on the one hand herbal plants have their basis of long-term application practice. On the other hand modern scientific knowledge and technologies have revealed that some new phytochemicals from natural plants could be beneficial for human beings.

In the foreseeable future, more and more herbal plants will cause interest and doubtlessly provide human beings with valuable agents of potential use in the research, prevention and treatment of various diseases and health problems.

One example is the plant Perilla, which has been traditionally used in Asian countries for medicine, garnish, food and food pigments, but is relatively unknown in the West. In recent years increasing literature has been published about the research and development of Perilla.

This book collects representative work about the plant Perilla, from its traditional use to newly found usage, from classical practice to modern scientific research, from cultivation to investigation, isolation and structural elucidation of some phytochemicals, from research to industrial development, from recent advances to future perspective of Perilla. Several of the chapters in this book touch on the application of Perilla extract for the treatment of allergy, the most widespread immunological disorder in humans.

It is our sincerest hope that this book will provide you with information on the past and present of Perilla that will help you to exploit its future.

He-Ci Yu
Kenichi Kosuna
Megumi Haga

ACKNOWLEDGEMENTS

Just as Perilla originated in Asian countries, so the contributors of this book are also mainly from Asia. I thank our invited contributors from China, Taiwan, Japan, Korea and also Finland for their excellent cooperation in preparing this book. I express my special thanks to our Book Series Editor, Dr Roland Hardman for his kind help. I am very grateful to Mr Arno Latvus, Managing Director of Hankintatukku Natural Products Co. for his considerable support. I am also very appreciative of Professors Lin Jen Hsou, Hu Jun Hong, Liu Ci Jun, and Mr Tony Mishima, Sami Isoaho, Yu He Jie, Gu Ruo Qing, Yu Hong Gu, Jian Mu, Liu Zhong Yuan, Yu Hong Chen and my colleagues in the Shanghai Pharmaceutical Bureau and Hankintatukku Co. for their kind assistance and support in producing this book.

CONTRIBUTORS

Yuh-Pan Chen
Brion Research Institute of Taiwan
116, Chung-Ching South Road, Sec. 3
Taipei, Taiwan

Tomoyuki Fujita
Department of Applied Biological
Chemistry, College of Agriculture
Osaka Prefecture University,
1-1 Gakuen-cho, Sakai,
Osaka 593, Japan

Zhi-Zhong Gong
Faculty of Pharmaceutical Sciences
Laboratory of Molecular Biology and
Biotechnology in Research Center of
Medicinal Resources
Chiba University
Yayoi-cho 1-33, Inage-ku
Chiba 263, Japan

Megumi Haga
Amino Up Chemical Co., Ltd
High Tech Hill Shin-Ei
363-32, Shin-Ei,
Sapporo 004, Japan

Lucy Sun Hwang
Graduate Institute of Food Science and
Technology,
National Taiwan University
59, Lane 144, Keelung Rd. Sec. 4
Taipei, Taiwan

Kiyoshi Kameda
School of Life Studies
Sugiyama Jogakuen University
17-3 Hoshigaokamotomachi, Chikusa
Nagoya 464, Japan

Young-Shik Kim
Department of Horticultural Science
Sangmyung University
Chonan 330-180, Korea

Tadao Kondo
Chemical Instrument Center
Nagoya University, Furo-cho, Chikusa
Nagoya 464-01, Japan

Kenichi Kosuna
Amino Up Chemical Co., Ltd
High Tech Hill Shin-Ei,
363-32, Shin-Ei,
Sapporo 004, Japan

Mitsuru Nakayama
Department of Applied Biological
Chemistry, College of Agriculture
Osaka Prefecture University,
1-1 Gakuen-cho, Sakai,
Osaka 593, Japan

Aimo Niskanen
Hankintatukku Natural Health Products
Co., Temppelikatu 3-5 A5,
Helsinki, Finland 00100

Kazuhiko Oyanagi
Odori Children's Clinic
Medical Building 3M
Odori Nishi 16-1, Sapporo,
Hokkaido; 060, Japan

Jukka Paananen
Hankintatukku Natural Health Products
Co., Temppelikatu 3-5 A5
Helsinki, Finland 00100

Kazuki Saito
Faculty of Pharmaceutical Sciences
Laboratory of Molecular Biology and
Biotechnology in Research Center of
Medicinal Resources
Chiba University
Yayoi-cho 1-33, Inage-ku
Chiba 263, Japan

Hyo-Sun Shin
Department of Food Science and
Technology, Dongguk University
26, 3-G, Pil-Dong, Jung-Ku,
Seoul 100-715, Korea

Mamoru Tabata
Professor Emeritus, Kyoto University
Sakuragaoka W8-17-5, Sanyo-cho
Okayama 709-08, Japan

Koji Tanaka
Showa College of Pharmaceutical
Sciences, Yakuyo Shokubutsuen
3-3165 Higashi Tamagawa Gakuen
Machida, Tokyo 194, Japan

Hiroshi Ueda
Department of Medicinal Chemistry
Faculty of Pharmaceutical Sciences
Teiko University, Sagamiko
Kanagawa 199-01, Japan

Masatoshi Yamazaki
Department of Medicinal Chemistry
Faculty of Pharmaceutical Science
Teikyo University, Sagamiko
Kanagawa 199-01, Japan

Mami Yamazaki
Faculty of Pharmaceutical Sciences
Laboratory of Molecular Biology and
Biotechnology in Research Center of
Medicinal Resources
Chiba University
Yayoi-cho 1-33, Inage-ku
Chiba 263, Japan

Kumi Yoshida
School of Life Studies
Sugiyama Jogakuen University
17-3 Hoshigaokamotomachi, Chikusa
Nagoya 464, Japan

Toshiomi Yoshida
International Center of Cooperative
Research in Biotechnology
Osaka University, 2-1 Yamada-oka
Suita-shi, Osaka 565, Japan

He-Ci Yu
Hankintatukku Natural Health Products
Co., Temppelikatu 3-5, A5,
Helsinki, Finland, 00100

Jian-Jiang Zhong
East China University of Science and
Technology, Research Institute of
Biochemical Engineering,
130 Meilong Rd., Shanghai 200237
P. R. of China

1.   INTRODUCTION

HE-CI YU

Hankintatukku Natural Products Co.
Temppelikatu 3-5 A 5, Helsinki, Finland 00100

I clearly remember in my childhood that, whenever we cooked crab or some other sea food, my parents would ask me to go to the Chinese medicine store to buy some herb leaves which was just what this book will describe—Perilla (*Perilla frutescens* Britt.), family Labiatae (Lamiaceae).

Like all knowledge of other traditional Chinese herbs, passed on from generation to generation, Perilla has its special effects in preventing and treating some diseases and health trouble. In Chinese folklore, Perilla leaf was known to prevent and neutralize the so called "poisoning" existing in crab or some sea food. In addition, a decoction of Perilla leaves was also drunk when some ailments such as a cold, cough, or indigestion occurred. In Japan, Korea and India, people commonly use Perilla leaves in the preparation of raw fish, and shell fish or grilled red meat. In Japan the purple-leafed Perilla is also used in the preparation of pickled plums. Perilla leaf is used both as a culinary herb and as a herbal medicine.

In the early 1990s some publications from Japan reported that Perilla leaf extract and Perilla oil had the ability to regulate the immune system and could be used as a health food in the treatment of some allergic reactions. A boom in the use of Perilla extract for allergy occurred in Japan. The efficacy of Perilla for allergy was widely recommended through various mass media to the public. Allergic patients experienced a satisfying result after self treatment with Perilla extract.

All this old and new knowledge of Perilla has created a great interest in the further understanding of the research work and application of Perilla. In fact, Perilla has a long history of benefit to human health in China and some other Asian countries. However, so far most publications of Perilla have been in Chinese, Japanese or Korean, relatively few publications exist in English. On the other hand, Perilla has recently received increasing attention. Searching the Medline™ database and CAB (Center for Agriculture and Biosciences International) database for "Perilla" shows a significant increase in the number of publications dealing with Perilla in recent years (Table 1).

DISTRIBUTION OF PERILLA

Perilla, an annual short-day plant is also called Wild Coleus, Beefsteak Plant (Duke, 1988), Purple Mint Plant and Perilla Mint (Wilson *et al.*, 1977) and Chinese basil. The scientific nomenclature of *Perilla* is confusing and has been discussed earlier by Zeevaart (1969,1986). It has been variously noted as *Perilla frutescens, Perilla arguta,* or *Perilla ocymoides* (Roecklein and Leung, 1987). Usually, two types of *Perilla* can be distinguished: the green-leafed varieties and the purple-leafed ones. Each type includes several forms.

Table 1 Numbers of Perilla Publications Searched by CAB and Medline

Database	Years	Numbers	Number per year
CAB	1987–1989	38	12.7
	1990–1992	58	19.3
	1993–1994	64	32
Medline	1976–1983	3	0.375
	1984–1990	24	3.4
	1991–1995	31	6.2

The green *Perilla* varieties have been described as:
P. frutescens (L.) Britt.,
P. frutescens (L.) Britt. var. *acuta* Kudo forma *viridis* Makino,
P. frutescens (L.) Britt. var. *crispa* Decne. forma *viridis* Makino,
P. frutescens (L.) Britt. var. *arguta* (Benth.) Hand.-Mazz.,
P. frutescens var. *acuta* f. *albiflora*,
P. frutescens var. *stricta* f. *viridifolia*,
P. ocymoides L.,
P. crispa (Thunb.) Tanaka var. *ocymoides* L.

The purple-leafed varieties have been described as:
P. frutescens (L.) Britt.var. *acuta* Kudo,
P. frutescens (L.) Britt. var. *typica* Makino,
P. frutescens var. *stricta*,
P. frutescens Britt.var. *crisp*,
P. frutescens var. *atropurpurea*,
P. frutescens (L.) Britt. var. *crispa* Decne. forma *purpurea* Makino,
P. ocymoides (L.) var. *nankinensis* (Lour.) Voss,
P. ocimoides (L.) var. *typica*,
P. crispa (Thunb.) Tanaka,
P. nankinensis (Lour.) Decne.

In the Japanese and Chinese literature some additional varieties or names occur:
P. frutescens (L.) Britt. var. *japonica* Hara (egoma),
P. frutescens (L.) Britt. var. *japonica* Hara fatropurpurea
P. frutescens (L.) Britt. var. *citriodora* Ohwi (remon-egoma),
P. frutescens (L.) Britt. var. *crispa* Decne. f. *discolor* Makino,
P. frutescens (L.) Britt. var. *crispa* Decne. f. *hirtella* Makino,
P. frutescens (L.) Britt. var. *crispa* (Thunb.) Handd.-Mazz.,
P. frutescens (L.) Britt. var. *crispa* (Thunb.) Handd.-Mazz. f. *atropurpurea*,
P. frutescens Britt.var. *acuta* Kudo f. *crispidiscolor* Makino,
P. frutescens L. var. *oleifera*.

Perilla is grown mainly in Asia and is native to the mountainous areas of China and India, being found up to an altitude of 1200 m (Roecklein and Leung, 1987). The major producing countries are China, India, Japan and Korea (Axtell and Fairman, 1992). In Southern Asia, it spreads widely from Kashmir to Bhutan and extends from Champaran to Burma (Misra and Husain, 1987). In its country of origin—China, Perilla is distributed over a wide area from 42 degrees of northern latitude southward and includes the Provinces of Hebei, Henan, Shanxi, Shandong, Jiangsu, Zhejiang, Jiangxi, Fujian, Hubei, Guangdong, Guangxi, Yunnan, Guizhou, Sichuan and also Taiwan. Among these areas, a large quantity of Perilla comes from Hubei, Henan, Sichuan, Shandong and Jiangsu. The better quality is mainly from Guangdong, Guangxi, Hebei and Hubei (Institute of Medicinal Plant Development, Chinese Academy of Medical Science, 1988; Zhao et al.,1993). The cultivation of Perilla in China were noted in Chapter 2 of this book (see also: Institute of Medicinal Plant Development, Chinese Academy of Medical Science, 1991) and is also described by Prof. Hwang in Chapter 14.

Perilla was introduced into Japan from China in 8th–9th century and is now grown extensively (Tanaka, 1993), its cultivation in Japan being described in Chapter 2. It is reported that in Japan good quality of Perilla comes from the north, in the Hokkaido area (Kosuna, Chapter 8).

In recent years, many studies on the growth conditions and quality improvement of Perilla have been reported by Korean scientists. During the period from January 1993 to August 1995, among the 75 articles found with Perilla as the key word covered by AGRIS database, more than half (40 articles) were published by Korean scientists. These articles included the improvement of Perilla cultivation and its seed oil quality (Lee et al., 1993a, 1993b; Lee et al., 1991a; Park et al., 1993). Lee et al. (1991b) reported a new high quality and high yielding Perilla variety "Okdongdlggae". It was developed by pure line selection and was released in 1990. Okdongdlggae produces dark grey seeds, and more branches and clusters per plant than the original variant Suwon 8. The leaves have an essential oil content of 0.36%. The seeds have a 44.8% oil content. It gave a mean seed yield of 118 kg/10 are, exceeding Suwon 8 by 12%. In South Korea 40,800 ha were in production in 1991 (Brenner, 1995).

In other parts of the world Perilla is less known and there is no appreciable acreage. Commercial plantings of Perilla in the United States, Cyprus, and South Africa have met with little economic success. Plants are found in the eastern United States as weeds (Roecklein and Leung, 1987). Mabberley (1987) noted that Perilla was also naturalized in the Ukraine and was also cultivated in South-Eastern Europe. It was not until recently that Perilla was first introduced into Finland along with the application of Perilla extract for the treatment of allergy. Cultural experiments of Perilla under Nordic climatic conditions have been made on a pilot scale in the area of Mikkeli located in northern latitude 61 degrees by a research group led by Dr. Galambosi of the Agricultural Research Center there. Cultivation was successful using black plastic 'mulch', which also served to remove the competition from weeds for an "organic growth" crop (Galambosi, Personal communication).

USES OF PERILLA

Medical Usage

As a raw material of herbal medicine, the leaf, stem and seed of Perilla has been used. In the recently published Chinese Pharmacopoeia (The Pharmacopoeia Commission of PRC, 1992), Perilla leaf, stem and seed were separately listed as traditional Chinese medicine for different purposes and a number of prescriptions containing Perilla leaf, stem or seed as one of the ingredients were collected. In Chapter 4 of this book Prof. Chen has presented a detailed review of Perilla prescriptions and their clinical application in ancient and modern China. In Chapter 10, Dr. Fujita and Prof. Nakayama have described some biological activities of constituents from Perilla. Prof. Tabata in Chapter 11 describes the pharmacology of Perilla involved in historical background and some results from modern scientific work. Some of these works supported the traditional medical application of Perilla, whereas some other works have revealed new findings for the application of Perilla.

In China some newer medical uses of Perilla have also been studied and developed such as the staunching action of Perilla leaf have been reported (Cao *et al.*, 1988) and the external use of Perilla leaf for the treatment of verruca (Yin and Guo, 1994). In India, Perilla (Bhanjira) is administered for cough and lung infections. It is used as a sedative, antispasmodic and diaphoretic, and is used in cephalic and uterine troubles (Badola *et al.*,1993).

Earlier, some patents were published on the inducing activity of interferon by Perilla leaf extract (Kitasato Institute, 1981, 1982, 1983). Recently, the new use of Perilla in the treatment of allergy has been developed by Japanese, Chinese and Finnish scientists. Prof. Yamazaki is one of the leading scientists in the studies on the immune regulating activity of Perilla extract. In Chapter 5 Prof. Yamazaki reports that water extracts of Perilla leaves rather than perillaldehyde could inhibit TNF (tumor necrosis factor) production both *in vivo* and *in vitro*. TNF overproduction is associated with acute or chronic inflammation. Oral administration of Perilla extracts have been shown to suppress chemical-induced inflammation and allergy. Therefore, Perilla extract might be a new candidate for an anti-inflammatory and antiallergic reagent.

In Chapter 6, a review of the allergy problem and the possibility of Perilla (Perilla extract and Perilla seed oil) as a new approach for the treatment of allergy is presented. As a health product, Perilla's usefulness in the relief of allergy symptoms is shown in the clinical trial reports from physicians and patients, and the investigations from the self-evaluation of patients. However, a systematic clinical trial data is still needed.

Dr. Oyanagi reports in Chapter 7, on the efficacy and experience of Perilla extract cream for the treatment of children with atopic dermatitis. He concludes that Perilla extract cream could be used with an efficacy rate around 80% in the treatment of some allergy without side effects.

In Chapter 8, Mr. Kosuna, the leading scientist in producing Perilla extract in Japan, describes the production process, the quality and the safety of Perilla extract. The active components showing TNF inhibiting activity are discussed. Not only the liquid preparation, but also the powder, capsule and topical cream of Perilla are used for ease of treatment.

Perilla Seed, Seed Oil and Essential Oil

As mentioned above Perilla seed is also used in herbal medicine. In India the seeds are often used in the curry materials when cooking meat, fish and vegetables (Misra and Husain, 1987). In Korea the seed and its oil are often used in the diet. The average seed yield in Korea is 770 kg/ha in commercial production (Brenner, 1995).The roasted seed is used widely as flavouring and as a rich source of nutrients.

The seed oil is similar to linseed oil in odour and taste but higher in drying quality. It is one of the richest vegetable sources of α-linolenic acid (n-3 fatty acid) and the Food and Agriculture Organization of The United Nations has classified Perilla as a minor oil crop (Axtell and Fairman, 1992). The oil is excellent for quick-drying paints and when incorporated with lac yields a brilliant and translucent yellow varnish. It has been substituted for linseed oil and tung-oil in paints, varnishes, linoleum, oil cloths, waterproof paper, printing inks, and similar items, showing excellent durability and high quality. The oil can also be used as raw material in the preparation of lacquers for leather. The content of seed oil varies from 29% to 52% depending on the variety and the commercial extraction rate of Perilla oil is approximately 40%. The expressed oilseed cake could provide a protein and fibre-rich livestock feed (Roecklein and Leung, 1987).

Studies on the lipid content and lipid composition of Perilla seed and seed oil of some different varieties have been carried out and reviewed by Prof. Shin in Chapter 9 of this book. He also describes the stability, nutritional and physiological value of the seed oil. Its α-linolenic acid (n-3 fatty acid) is closely associated with anti-hypertensive effect, antithrombosis effects, antiallergic action and inhibition in carcinogenecity.

In addition, Perilla yields commercially important aromatic volatile oil from its leaves and flowering tops. Perilla contains 0.3–1.3% (dry weight) of a natural aromatic oil. The content is highest when the flowers first appear. The leaves contain 1.04% and the stems 0.05% (by dry wt.) and production involves steam distillation (Chemical Abstracts, **51**, 8380f).

Prof. Tabata (Chapter 11, this book) and Dr. Fujita and Prof. Nakayama (Chapter 10) indicate the various chemical constituents isolated from essential oil of Perilla and their biological and pharmacological activities. So far six chemotypes which are genetically stable have been detected. Chinese scientists (Zhao *et al.*, 1993) have also grouped *Perilla frutescens* into three different chemotypes on the basis of plant morphology and the composition of the leaf essential oils.

The first report of the ketone, rosefuran, comprising the major component (58%) of an essential oil came from Perilla flower oil from a Bangladeshi strain of *P. ocimoides (P. frutescens)* grown at Lucknow (Misra and Husain, 1987). Rosefuran contributes to the characteristic odour of Bulgarian rose oil (and now produced in India). Having in mind the high price of rose oil, Perilla oil may be a commercial source of rosefuran for use in perfumery.

RECENT STUDIES ON PERILLA

Perilla frutescens and related species have also been used extensively in plant biochemistry and physiology studies (Danilova and Kashina, 1995; Zeevaart,1986).

In Chapter 12, Dr. Mami Yamazaki *et al.* briefly review on genetic studies of Perilla and report their preliminary work on genes for anthocyanin biosynthesis in Perilla.

In Chapter 10 Dr. Fujita and Prof. Nakayama describe the chemical structures and the biological activities of more than 60 active constituents presented in the Perilla plant. The intact Perilla plant contains a large amount of phytochemicals including non-volatile components (terpenoids, sterols, phenolics, cinnamates, phenylpropanoids, flavonoids, anthocyanins and their glycosides) and volatile components (mono- and sesquiterpenes, and some minor components).

In Chapter 3, Dr. Zhong from China and Prof. Toshiomi Yoshida from Japan have collaborated well in establishing the technique of cell culture of Perilla. They present a review and describe a systematic study on the optimisation of culture conditions, biological, physical and chemical factors for anthocyanins production in the culture. It is interesting that they indicate the future possibility of commercial anthocyanin production with high yield by cell cultures.

In order to obtain a particular phytochemical group, changing the culture method is an effective way of securing some compounds. Perilla callus tissue was found to contain almost exclusively caffeic acid derivatives as the phenolic extractives. It is easier to isolate caffeic and rosmarinic acids from the Perilla cell culture rather than from the intact plants. Thus, cell suspension cultures could be effectively used for medicinal purposes as a dry cell powder or as an alcoholic cell extract containing caffeate at a high rate. Furthermore they could be constantly supplied to the market throughout the seasons (Ishikura, 1991). Caffeic and rosmarinic acids are useful secondary metabolites as an antiphlogistic agent and for other remedies. Mr. Kosuna indicates (Chapter 8, this book) that caffeic and rosmarinic acids are among the active compounds in the Perilla extract which showed antiallergic action.

The use of natural anthocyanins as safe food colorants, antioxidants and free radical scavengers has become popular in recent years. In Chapter 13 Dr. Kumi Yoshida *et al.* present a review of Perilla anthocyanins and flavones from the chemical view point and in Chapter 14, Prof. Hwang mainly describes the isolation of anthocyanins by membrane technique. Both have investigated the control of colour stability, important in the practical application of these compounds as natural colorants.

REFERENCES

Axtell, B.L. and Fairman, R.M. (1992) *FAO Agricultural Services Bulletin* No. 94, 107.

Badola, K.C., Pal, M. and Bhandari, H.C.S. (1993) Effect of auxins on rooting shoot cuttings, growth and flowering of Bhanjira (*Perilla frutescens* Linn.). *Indian Forester* **119**, 568–571.

Brenner, D.M. (1995) Perilla 1–6.
in: http://newcrop.hort.purdue.edu/hort...s/Crops/CropFactSheets/perilla.html

Cao Y., Zhao Z.W., Yang Y., Zhang X.P., Zhu N.J., Wang Y. (1988) Efficacy of Perilla for treatment of hysterotrache bleeding. *Zhong Yi Zhi,* **1988**, No. 8, 49 (in Chinese).

Danilova, M.F. and Kashina, T.K. (1995) Photoperiodism, leaf development and the dimorphism of chloroplast thylakoids in *Perilla ocymoides* L. *Russian J. of Plant Physiology*, **42**, 9–16 (in Russian), (English Abstract).

Duke, J.A. (1988) *Handbook of Medicinal Herbs*, CRC Press, Boca Raton, FL, pp. 354.

Institute of Medicinal Plant Development, Chinese Academy of Medical Science (1988) *Zhong Yao Zhi,* **1988**, No. 4, 668 (in Chinese).

Institute of Medicinal Plant Development, Chinese Academy of Medical Science (eds.) (1991) *Horticulture of Chinese Medicinal Plants*, Agriculture Publishers of China, pp. 1077–1078 (in Chinese).

Ishikura, N. (1991) *Biotechnology in Agriculture and Forestry, vol. 15 Medicinal and Aromatic Plants III* (ed. Y. P. S. Bajaj) p. 353–361.

Kitasato Institute (1981) Pharmaceuticals containing interferon inducers from *Perilla frutescens* crispa. *Jpn. Kokai Tokkyo Koho JP.,* **81,** 166119.

Kitasato Institute (1982) Interferon inducers from *Perilla frutescens crispa. Jpn. Kokai Tokkyo Koho JP.* **82**, 131724.

Kitasato Institute (1983) Interferon inducers from Perilla leaves and stems. *US Patent* 4419349.

Lee, J.I., Bang, J.K., Lee, B.H., Kim, K.H. (1991a) Quality improvement in Perilla-(1) Varietal differences of oil content and fatty acid composition. *Korean J. of Crop Science (Korea Republic),* **36**, 48–61(in Korean), (English Abstract).

Lee, J.I., Han, E.D., Bang, J.K., Park, H.W., Park, R.K. and Park, N.J. (1991b) A new high quality and high yielding Perilla variety "Okdongdlggae". Research Reports of the Rural Development Administration, Upland and Industrial Crops **33**, 49–53 (in Korean), (English Abstract).

Lee, J. I., Park, C.B. Son, S.Y., (1993a) Quality improvement in *Perilla*-(3) Varietal differences of protein content and amino acid composition in *Perilla. Korean J. of Crop Science (Korea Republic),* **38**, 15–22 (in Korean), (English Abstract).

Lee, J. I., Park, C.B., Lee, B. H., Kang, C. W., Cho, C. Y., Park, R.K. (1993b) A new *Perilla* variety "Daeyeupdlggae" characterized by large seed and leaf with high quality. *RDA-J. of Agricultural Science (Korea Republic)* **35**, 185–189 (in Korean), (English Abstract).

Mabberley, D. J. (ed.) (1987) *The Plant Book.* pp. 442. Cambridge University Press.

Misra, L.N. and Husain, A. (1987) The essential oil of *Perilla ocimoides*: a rich source of rosefuran. *Planta Med.* **1987**, 379–380.

Park, C. B.; Lee, J.I., Lee, B.H.; Son, S.Y. (1993) Quality improvement in Perilla -(2) Variation of fatty acid composition in M2 population. *Korean J. of Breeding* (in Korean), (English Abstract).

Roecklein, J. C. and Leung, P.S.(eds.) (1987) *A Profile of Economic Plants*, p. 349, New Brunswick N.J.

Tanaka, K. Effects of *Perilla* (1993) *My Health,* **1993**, No.8 152–153 (in Japanese).

The Pharmacopoeia Commission of PRC (1992) *Parmacopoeia of The People's Republic of China* p. 18, 55, 78.

Wilson, B.J., Garst, J.E., Linnabary, R.D. and Channell, R.B. (1977) Perilla ketone: a potent lung toxin from the mint plant, Perilla frutescens Britton. *Science*, **197**, 573–574.

Yin Jian and Guo Li-gong (eds.) (1994) *Modern Research and Clinical Application of Chinese Medicines*. pp. 632–633. (in Chinese)

Zeevaart, J.A.D.(1969) Perilla, in *The induction of flowering. Some Case Histories*, Evans, L.T., ed., Macmillan of Australia, Melbourne, 116–155.

Zeevaart, J.A.D. (1986) Perilla, in *CRC Handbook of Flowering*, Vol. 5 p. 239–252.(ed. Abraham H. Halery, CRC Press).

Zhao S., Zhu Z., Feng Y., Quan L. and Xue L. (1993) Studies on chemical components of essential oil of different chemical type from *Perilla frutescens* (L.) Britt. var. *argata*. and *P. frutescens*. *Natural Product Research and Development* , **5** (3), 8–12 (in Chinese), (English Abstract).

2.   CULTIVATION OF PERILLA

KOJI TANAKA[1], Y. S. KIM[2] and HE-CI YU[3]

[1]Medicinal Plant Garden, Showa College of Pharmaceutical Sciences, 3-3165 Higashi
Tamagawa Gakuen, Machida, Tokyo 194, Japan
[2]Department of Horticultural Science, Sangmyung University, Chonan 330-180, Korea
[3]Hankintatukku Natural Products Co., Temppelikatu 3-5 A 5, Helsinki,
Finland 00100

Perilla (*Perilla frutescens* Britt. family Labiatae), is an annual short-day plant which originated in China and India. It was introduced to Japan in about the 7–8 century and now is extensively grown there and in Korea where it has a long history also. Currently Korea could have the largest area of cultivation of Perilla. This was 49,900 ha in 1992 (Table 1).

FORM DESCRIPTION

In Japanese, Perilla is called Shi So (or Ji So) which was derived from the pronunciation of the Chinese name Zi Su (or Tzu Su) which refers to the purple colour (Zi) of its leaf and Su referring to its soothing and recuperative properties after some illness. In German and French it is named Schwarznessel and Perilla de Nankin, respectively.

The plant, with a height of 0.5–1.5 m, superficially resembles basil and coleus (Figure 1).

Figure 1   Green Perilla

Table 1. The Cultural Area and Seed Production of Perilla in South Korea*

Production	1980	1985	1990	1991	1992	1993
Cultural area (ha)	21,700	28,100	37,100	40,700	49,900	41,200
Average productivity (kg/ha)	560	740	760	740	740	690
Nation-wide production (1000 ton)	12.1	20.7	28.0	30.2	36.8	28.4
Total sales value (million US dollar)	9.0	26.8	62.9	84.6	117.1	unknown

*Ministry of Agriculture of Korea, 1994

Stem and Leaf

The stem (0.5–1.5 cm in diameter) is branching and brown-purple or dark purple in colour. Its cross-section is near square with four obtuse angles. The leaves are oval (4–12 cm long and 2.5–10 cm wide) with an acumen and dentate edge. Leaves grow opposite with a petiole 2–7 cm long. The surface of stem and leaf is somewhat pubescent and the hairs give rise to the unique pungent odour and taste.

Flower

The inflorescence is axillary with terminal raceme 6–20 cm long. The purple or white flower is small with pubescent campanulate calyx and lipped corolla. The corolla has five lobes, two upper and three lower. Flowering is from June to August according to the location planted.

Fruit and Seed

Fruiting time is July to October and the ripe fruit is a collection of greyish-brown nutlets containing 1–4 granules of seed. The seeds are ovoid or subspherical, about 0.6–2 mm in diameter, greyish-brown to blackish-brown, and with a net-patterned testa. The seeds have a slightly pungent taste.

VARIETIES AND USE OF PERILLA IN JAPAN

Perilla generally occurs as purple or green Perilla according to the plant colour. They have been used for folk medicines, diet or garnish, which varies according to the varieties, growth stage and the parts used: leaf bud (actually a root free seedling plant), leaf, stem, flower head and seed.

In Japan the Perilla which is used for its leaf bud is called "Mejiso" (bud Perilla). Similarly they have "Hajiso" (leaf Perilla), "Hojiso" (head Perilla) and "Shiso-no-mi" (fruit Perilla). Bud Perilla has been used as spice when eating raw fish. In recent years the purple bud has been mainly used and cultivated. The flower heads of Perilla, with about 30% of the flowers open, are harvested and also used as spice when eating raw fish. While the head Perilla, in which some seeds are already formed in the flower head, is used with fried fish or shrimp (tempura) or with some Japanese dishes made of soy sauce, sugar etc. The mature fruit is used in the preparation of processed foods such as pickles etc.

The following are cultivated and are used mainly as commercial sources of bud Perilla (purple or green), "Hajiso", "Hanajiso", "Hojiso" and "Shisonomi".

Perilla frutescens (L.) Britt. var. *acuta* (Thunb.) Kudo ("Aka Shiso" in Japanese)

The leaves are purple and the flowers are pale purple. It is mainly used as "Mejiso" and also for its purple flower heads.

Perilla frutescens (L.) Britt. var. *acuta* Kudo f. *crispidiscolor* Makino ("Katamen Shiso")

Leaves are green on the obverse side and purple striped on the reverse side. It has stronger fragrance and is mainly used as "Mejiso" and "Hajiso".

Perilla frutescens (L.) Britt. var. *crispa* (Thunb.) Hand.-Mazz. f. *atropurpurea* ("Aka Chirimen Shiso")

Its stem is reddish purple. The obverse side of the leaf is dark purple and the reverse side is light purple. Leaves are big, soft and with some wrinkles. Its flower is pale purple and late maturing. It is mainly used as "Mejiso" and "Hajiso".

"Wase Chirimen Shiso"

The plant is smaller and grows faster than "Aka Chirimen Shiso". It is mainly used as "Mejiso".

Perilla frutescens (L.) Britt. var. *crispa* (Thunb.) Hand.-Mazz. ("Ao Chirimen Shiso")

The leaves are green on both sides. The leaf is big and with many wrinkles and the flower is white. It is mainly used as "Mejiso", "Hajiso", "hojiso" and "Shisonomi".

CULTIVATION

Here are described some cultivation techniques from different areas.

Cultivation of Perilla in Japan

It was reported in Japan that the total acreage of Perilla in 1995 was about 103,000 ace (one ace=100 m^2) of which about one tenth of that was in the Hokkaido area (Megumi Haga, Personal Communication).

Seed

The shelf life of seeds is about one year. In Japan seeds are stored in the refrigerator (0°–3° C) after May and throughout the summer time. The refrigerator stored seeds are kept at room temperature 4–5 days before sowing. Gibberellin may be used to break dormancy. One litre gibberellin solution (50 ppm) is added to 1.8 litres seed and set aside until all the solution has been absorbed. Then the seed is sown. Germination occurs in 6–10 days, the optimal temperature being 22° C. Dried seed absorbs water with difficulty. To avoid excessive drying during storage the seed may be mixed with sand or soil and placed in a bag or box underground.

Conditions

The optimal temperature for growth is about 20° C. However, the crop can tolerate a higher temperature. It is easily cultivated in most soils other than extremely dry.

Cultivating "Mejiso" (bud Perilla)

The seed bed is a raised bed 1.2–1.5 m wide with manured soil or fertile soil mixed with plant derived ash. The seed is soaked in water for two days, the water being replaced once. The seed is spread evenly over the seed bed at the rate of 9 ml per square meter. Then the bed is covered with a layer of river sand and pressed down lightly using a board. After fully watering, the beds are covered with gauze and a vinyl tunnel in order to keep the beds humid. Before germination it is important to keep humidity very high. Afterwards the humidity is reduced and the seedlings are given enough sunlight to colour the cotyledon.

   The best time to harvest green bud is when the cotyledons are expanded and the first true leaf has grown out. It takes 7–10 days in summer or 15–20 days in the other seasons. For the purple bud, it is harvested when two true leaves have grown out. So it takes about twice as long for purple bud as for the green bud. For harvesting it is better to use scissors or a thin knife and cut the buds with hypocotyls (4–5 mm). After washing with water the buds are placed in wooden boxes each of capacity 100 ml. The yield is 20 boxes per square meter.

Cultivating "Hojiso" (head Perilla)

Seed, 100–200 ml per ace, is sown in a warm or cool bed. After 30–40 days the plants are transplanted and then planted a second time 20 days later. The field used should have received a basal dressing of compost followed by a fertiliser giving 1.5 kg nitrogen (N), 2 kg of phosphorus (P) and 2 kg of potassium (K) per ace.

   Under long day conditions shorten the day light time to 7 hours by making shade for the plant. With short days use artificial light for up to 16 hours.

   When five or six flowers are opening, the crop is harvested by cutting the flower head 1.5 cm below the head and tidily packing 30 heads per box.

Table 2   Procedure for Perilla cultivation in Hokkaido

Sowing Time	End of April–early May.
Planting Field	Any kind of soil except for the field of natural growth and also the field in which Perilla was cultivated the previous year.
Planting Density	800 plants/ace, row width–60 cm, spacing–20–25 cm.
Sowing	30 ml of seed per acre. 68,000 seeds weigh 55 g (100 ml). Sow when the soil contains enough moisture, lightly cover the seed with the soil and press it down carefully.
Fertiliser	Fertiliser standard (per acre): N 1 kg, P 0.65 kg and K 0.66 kg.
Weeding	Middle of June–Middle of July. Remove the weeds before as they grow too thick when the weather is favourable. Tall weeds must be removed.
Thinning	Thinning should be started after the 4th or 5th leaf has appeared and should be completed before the plants has reached a height of 15 cm.
Supplementary Sowing	If the germination is poor, supplementary sowing is carried out on vacant hills.
Pest Control	Chemicals used to control insects (such as striated chafer, aphid, spider mite, and cabbage army worm) must not be applied one month before harvesting.
Harvesting	Hand or machine cutting applied so as to obtain as much as possible.
Drying	Dry the leaves in the sun to a moisture content of about 13%.

Cultivation of Perilla in Hokkaido Japan

Table 2 lists the standard procedure for the Perilla cultivation in Hokkaido, where the crop is mainly used for the production of leaf extract (Kosuna Kenichi, Personal Communication).

Cultivation of Perilla in Korea

In Korea Perilla (green leaf) has been traditionally used for its seed oil with early and late maturing oil seed cultivars. Nation-wide production scale of seeds is about 36,800 ton/year and average productivity is 740 kg/ha in 1992. Perilla leaves have been obtained as a by-product and consumed in a salted form or a wrapped form with meat and/or fish. Leaf consumption tends to increase with the rise in the standard of living. Perilla is a warm-season crop and sensitive to daylength (Yu, 1974; Choi, 1980; Lim, 1989). Hence production falls in winter when consumption is at its highest and this is reflected in the leaf price 2.5 US$/kg in winter and 1 US$/kg in summer.

Cultivars

The representative cultivars are Suwon No.8 and No.10 for open field cultivation, and Guppo for winter cultivation. Many cultivars and ecotypes are also cultivated locally.

Table 3   Yields of *Perilla frutescens* leaves according to the planting density*

Planting density	40×10 cm	30×10 cm	20×10 cm
Yields (g/plant)[z]	2.49 a[y]	2.46 a	2.28 b
Yields(g/unit area)	7.47 c	9.84 b	13.68 a

[z] harvested from April 13 to May 4, 1993 (Nodes are 5 to 11).
[y] means with same letter within a row are not significantly different at the 5% level by Duncan's multiple range test.
*Kim, 1995

Climatic requirements

The photosynthetic rate of Perilla leaf is increased from 10° C to 25° C and does not change above 25° C. The temperature must be maintained over 10° C at night. Light saturation point is about 12,000 lux and light compensation point is about 5,000 lux. Perilla is known as a typical short-day plant. A "night-break" treatment is therefore needed for leaf growth. During the winter season growers often install 60 W incandescent lamps one meter above the plants (> 100 lux) for 1–2 hours after midnight to inhibit floral initiation.

Agronomy

1. Cropping systems: In open fields, seeds are sown from late April to early June. Leaves are harvested from mid June to early September. In a forcing system, sowing is October and harvesting in December to March. In retarded system, sowing takes place in August with harvesting in September to March.
2. Planting and spacing: The amount of seed used for planting is 3 kg/ha. Studies of planting density of Perilla have been done exclusively on soil culture, and are mainly concerned with oil production (Yoo and Oh, 1975; Mok et al., 1972). Yield is increased to a certain level by high density(Table 3). The limitation of density was 250,000 plants per ha (planting distance 40×10 cm) in summer when high density is adopted. Above this density the plants are too tall and thin (Lim, 1989). The high planting density is obtained in hot season. The factors affecting density limitation are light intensity, temperature, nutrition, water, and CO_2 concentration.
3. Fertiliser treatment: The general dressing before planting is recommended as follows: compost 1000 kg, N 4, P_2O_5 3, K_2O 2 kg/10a. Urea 70 and KCl 50 kg/ha are applied as top dressing after the first harvest and after the third harvest.

Hydroponic production

Hydroponics can overcome the limiting factors of soil culture such as supply of nutrients and water and reduce labour, in which planting density can be higher than that in soil culture. The leaf emergence rates are not affected by planting density. The rate of increase of leaf number becomes greater 46 days after germination; the leaf number reaches twenty, 91 days after germination (Kim, 1995). The area given over this production technique has increased markedly in the southern districts.

The method of production of seedlings depends on the type of hydroponic system and grower's experience. In water culture, urethane, rockwool or peat mixture blocks are often used. The seeds germinate in 4 to 10 days depending on the temperature. One third strength nutrient solution is supplied when the cotyledons are fully expanded and the concentration is increased with growth.

One month after sowing, the seedlings with 2–3 true leaves are transplanted to Styrofoam raised beds supplied with nutrient solution. Perilla prefers NH_4 to NO_3 nitrogen and a ratio of 3:9. is recommended (Kim, 1993). The composition of macronutrients is as follows: $NH_4H_2PO_4$ 114, $Ca(NO_3)_2$ $4H_2O$ 472, KNO_3 505, K_2SO_4 261, NaH_2PO_4 40, $MgSO_4 7H_2O$ 492 g/L (Kim, 1995). The composition of micronutrients are the same as Yamazaki's nutrient solution (Yamazaki, 1982). The pH and EC (electric conductivity) are controlled at 5.5–6.5.

Any kind of hydroponic system may be used for Perilla. In Korea either the nutrient film technique (NFT) or deep flow technique (DFT) is preferred, as used for other leafy vegetables. In these techniques the plants are secured so that the roots are bathed in a solution of nutrients. Root respiration is affected by the distance of the base of the stem from the solution with 5 to 8 cm being recommended, depending on the species and the environment conditions (Park and Kim, 1991). For Perilla 5 cm is preferred in the summer (Table 4) and this may remain appropriate for winter, because the dissolved oxygen is greater then and excessive exposure of the root to lower temperature is to be avoided. The system basically consists of cultured beds or channels, with a circulating pump, tank for nutrient solution, valves and lines. Nowadays the system is supplemented with filters, fertiliser tanks, sterilisers, etc, and is automatically controlled, the nutrient solution being circulated for about 15 minutes every hour and renewed every three weeks.

Table 4   Growth for solution-stem base distance in *Perilla frutescens*[*]

Treatment	Deep	Medium	Shallow
Plant height (cm)[z]	53.5	51.9	53.2
Base cir. (cm)[y]	3.1 b[x]	3.1 b	3.2 a
Root weight[y]			
fresh weight (g)	975 b	985 b	1360 a
dry weight (g)	38.5 b	40.0 ab	43.6 a
% dry weight	4.0 a	4.1 a	3.2 b
Yields (g/plant)[w]	28.6 b[y]	28.6 b	35.5 a

[z] measured on Dec. 28, 1992.
[y] measured on Jan.14, 1993.
[x] means with same letter within a row are not significantly different at the 5% level by
  Duncan's multiple range test.
[w] period of harvest is Nov. 7, 1992–Jan.7, 1993. Yields are the average of 12 plants.
[*] Kim, 1995.

Cultivation of Perilla in China (Institute of Medicinal Plant Development, Chinese Academy of Medical Science, 1991)

Fields

Perilla fields should be without shade and convenient for irrigation. The soil should be loose and fertile. The field is prepared in early April. After applying compost or barnyard manure (300–450 kg/ace) as ground fertiliser the fields are ploughed to a depth of 25 cm and raked level.

Cultivation

Seeds or transplanted seedlings are used. Sowing directly with seed gives faster growing plants and an earlier harvest. It also saves labour by avoiding the transplanting stage. However, when sowing with seed, it is important to plant for the optimum plant density, otherwise the crop yield will be adversely affected.

The seed is sown in mid or late April in Northern China and in late March in Southern China. Sowing may be in a row pattern or a hole pattern. In the former pattern the row distance is 30–50 cm and the depth is 0.5–2 cm. The amount of seeds is about 110 g/acre. For the hole pattern the hole could be 30×50 cm and the amount of seeds is 22 g/acre. Under the optimum temperature (about 25°C) it will take about 5 days for a seedling to be produced.

Cultivation *via* seedling is best suited to when the field is not yet available or when seed supply is limited, or in the absence of irrigation. Sowing time is as above, into an adequately watered and fertilised bed, the seed being covered with a layer (0.5 cm) of fine soil. The seedlings appear in 7–8 day. In early May, when the young plants are about 15–20 cm tall, they are transplanted. The previous day the bed is watered thoroughly to limit the damage to the roots. The field should be ready with trenches opened 15 cm deep and with a row distance of 50 cm. The planting distance in the rows is about 30 cm. After returning the soil the plants are well watered. Subsequently the crop is given less water to encourage the roots to enter the deep soil and absorb the nutrients for good growth.

Fields Management

Fertilising: Using fertiliser promotes good growth. If the soil is poor or without ground fertiliser the young plants are given chemical fertiliser 2–3 kg/acre every other week: 15–20 kg (containing 1.5 kg of N, 1.5 kg of P, and 1kg of K) for the whole growth time. If additional barnyard manure is used this should be during June to August, once a month at the rate of 225 kg/ace each time. The first application of manure should be small since the young plants are tender and after the last application the soil should be drawn up.

Irrigating and draining: Perilla needs more water as a young plant or when blooming. In the rainy season it is important to drain the water in good time.

Pest control and disease prevention: The usual insect pests are *Pyrausta phoeniccalis* Hubna, *Plusia* sp., *Agrotis ypsilon* Rottemberg and *Cryptaphis siniperillae* Zhang etc. and some diseases are *Septoria perillae* Miyake, *Cuscuta australis* R. Br., and rust. Measures for preventing diseases are: avoiding planting too densely, draining the water in the rainy season, removing and destroying diseased plants (and their seed) and using some chemicals.

Harvesting

The harvesting time varies with the intended use of the crop and the climate of the growing area. The amount of volatile oil is highest when the flower head has just grown out 1.5–3 cm. Therefore for this oil production, the harvest time is at the beginning of flowering in August to September in Shanghai area. Usually 225 kg sun dried leaf yield 0.2–0.25 kg essential oil. For the medicinal use of the leaf and stem, the harvest time is when the leaves and branches are flourishing from July to August in Southern China or from August to September in Northern China. When making use of the whole plant (leaf, stem and seed), harvesting time is usually from September to October. It was reported that the National Traditional Chinese Medicine Corporation of China collected over 5000 ton dried Perilla material in 1990 (National Traditional Chinese Medicine Corporation of China, 1994).

REFERENCES

Choi, G.H. (1980) Studies on the quantitative analysis of growth and response of photoperiods in Perilla (*Perilla ocymoides*). MS thesis, Hyosung Women's University, Korea (in Korean).
Institute of Medicinal Plant Development, Chinese Academy of Medical Science (eds.) (1991) *Horticulture of Chinese Medicinal Plants*, Agriculture Publishers of China, 1077–1078 (in Chinese).
Kim, Y.S. (1993) The effect of NO_3-N and NH_4-N ratio on the growth of *Perilla frutescens* in hydroponics. *J. Bio. Fac. Env.* **2**, 119–126 (in Korean).
Kim, Y. S. (1995) Studies on the distance from stem-base to solution, and the planting density for the growth of *Perilla frutescens* by deep flow culture. *Acta Hort.* **396**, 75–82.
Lim, C. I. (1989) Studies on the year-round cultivation of Perilla for leaf production. PhD thesis, Korea University, (in Korean).
Ministry of Agriculture of Korea (1994) *Statistical Yearbook of Agricultural Forestry and Fisheries.*
Mok, I. J., Lee, S. S. and Kim, Y.W. (1972) Experiment on the planting distance of Perilla. *Annual Res. Rep. Korean Hort. Exp. Sta.* **1972**, 361–368 (in Korean).
National Traditional Chinese Medicine Corporation of China (eds.) (1994) *Resource of Traditional Chinese Medicine* Science Publishers of China, 1096 (in Chinese).
Park, K. and Kim, Y.S. (1991) *Principles and practices in hydroponics.* Korea University Press, 121–123 (in Korean).
Yamazaki, K. (1982) Soilless culture. Hakuyu Press, Tokyo, (in Japanese).
Yoo, I. S. and Oh, S. K. (1975) Experiment on the relationship between planting density and fertilizer quantity. *Annual Res. Rep. Korean Crop Exp. Sta.* (Speciality Crop) **1975**, 118–123 (in Korean).
Yu, I. S. (1974) Studies on the responses to daylength and temperature and their effects on the yield of Perilla (*Perilla ocymoides*). *J. Korean Soc. Crop Sci.* **17**, 79–114 (in Korean).

3. CELL AND TISSUE CULTURES OF *PERILLA*

JIAN-JIANG ZHONG[1] and TOSHIOMI YOSHIDA[2]

[1]East China University of Science and Technology, State Key Laboratory of Bioreactor Engineering, 130 Meilong Road, Shanghai 200237, China
[2]International Centre for Biotechnology, Osaka University, 2-1 Yamada-oka, Suita-shi, Osaka 565, Japan

INTRODUCTION

In recent years, plant cell and tissue culture has received a great deal of attention for the production of useful plant-specific chemicals. The advantages of plant cell culture compared with whole plant cultivation are that: cell cultures are unlimited by environmental, ecological or climate conditions; cells can proliferate at a higher growth rate; a large amount of cells can be obtained in a bioreactor with limited space; metabolite accumulation can be improved through regulation of culture conditions; downstream processing of products from cell cultures may be relatively easier.

The first report dealing with cell and tissue cultures of *Perilla* may be that by Sugisawa and Ohnishi (1976), and there have been many related publications since the 1980s. As shown in Table 1, these works were concerned with the formation of perilla pigments, caffeic acid, monoterpenes, sesquiterpenes, and essential oil by *Perilla* cells, as well as glucosylation, resolution, and morphogenesis of the cell cultures.

The cultural factors which affect cell growth and metabolite accumulation in plant cell cultures include biological (cell line, culture age, inoculum density, and cell aggregate size), chemical (such as medium composition), and physical factors (such as temperature, light irradiation, oxygen supply, and shear stress).

CELL AND TISSUE CULTURES

Callus Culture

Basic requirements and media

The basic requirements for plant tissue culture work are: (i) an area for medium preparation; (ii) a sterile room or sterile air cabinet for aseptic transfer; (iii) a constant temperature room or incubator for growth of callus cultures; (iv) shaker facilities for cell suspension cultures. The main physical requirement for growth and maintenance of plant cell cultures is constant temperature. Callus cultures are grown in plastic Petri dishes, glass culture tubes or plastic pots with screw cap lids. Suspension cultures are usually in glass conical flasks (Dixon, 1985).

Components of media for the growth of plant callus and suspension cultures can be classified into six groups, and this division is usually reflected in the way in which stock

Table 1   Reports on cell and tissue cultures of *Perilla*

	Culture conditions	Authors (year)
Metabolites:		
perilla pigments	MS medium, 100 ppm NAA, 2 ppm KT, 25°C, with light	Ota (1986)
	LS medium, 10 μM NAA, 1 μM BA, 25°C, light 3000 lux for 12 h	Koda *et al.* (1992)
	LS medium, 1 μM 2,4-D and 1 μM BA, 25°C, light at 17–20.4 W/m²	Zhong *et al.* (1991, 1993a, 1994a)
phenylpropanoids	B₅ medium, 5 ppm NAA, 1 ppm KT, 25°C, light at 2000 lux	Tamura *et al.* (1989)
caffeic acid	MS medium, 1 ppm 2,4-D, 0.1 ppm KT	Ishikura *et al.* (1983)
monoterpenes	MS medium, 1 ppm 2,4-D, 5 ppm KT 25°C, slightly dark	Sugisawa & Ohnishi (1976)
sesquiterpene	MS medium, 1 ppm NAA, 1 ppm KT, 25°C, light 3000 lux	Nabeta *et al.* (1985)
	modified MS, 1 ppm 2,4-D, 5 ppm KT	Shin (1986)
ursolic acid	LS medium, 1 μM NAA, 10 μM KT	Tomita & Ikeshiro (1994)
cuparene	MS medium, 1 ppm NAA, 1 ppm KT 25°C, light at 3000 lux	Nabeta *et al.* (1984)
essential oil	modified MS, 1 ppm NAA, 5 ppm KT 27±2°C	Shin (1985)
Glucosylation	LS medium, 1 μM 2,4-D, 25°C, dark	Tabata *et al.* (1988)
	MS medium, 1 μM 2,4-D, 26°C, dark	Furukubo *et al.* (1989)
Resolution	LS medium, 2,4-D, 26°C, dark	Terada *et al.* (1989)
Morphogenesis	MS medium, NAA, or 2,4-D, BA, NOA	Tanimoto & Harada (1980)

Abbreviations: BA, benzylamino-purine; 2,4-D, 2,4-dichlorophenoxyacetic acid; KT, kinetin; NAA, 1-naphthaleneacetic acid; NOA, naphthoxyacetic acid. MS: Murashige and Skoog's, LS: Linsmaier and Skoog's.

solutions are prepared and stored. The groups are: (i) major inorganic nutrients; (ii) trace elements; (iii) iron source; (iv) organic supplement (vitamins); (v) carbon source; (vi) organic supplement (plant growth regulators). Table 2 shows a typical medium for *Perilla frutescens* cell cultures (Zhong *et al.*, 1991).

Callus induction

The callus of *P. frutescens* was induced as follows: seeds were germinated on an agar medium of LS (Linsmaier and Skoog) minus growth regulators to produce young seedlings for use as a source of explants. Young leaf sections (5 mm²) were excised and transferred to MS (Murashige and Skoog) basal medium containing sucrose (30 g/L), 2,4-dichlorophenoxyacetic acid (2,4-D) (1.0 ppm), kinetin (KT) (5.0 ppm) and Difco bacto-agar (0.9% w/v). The callus tissue was subcultured every 3 weeks during 6 months at 25°C in poor light conditions. A suspension culture derived from the callus of the

Table 2   Medium composition for cell cultures of *P. frutescens*

Inorganic	Concentration (mg/L)	Organics	Concentration (mg/L)
KNO$_3$	1900	Myo-inositol	100
NH$_4$NO$_3$	1650	Thiamine HCl	0.4
CaCl$_2$·2H$_2$O	440		
MgSO$_4$7H$_2$O	370	Sugar	Concentration (g/L)
KH$_2$PO$_4$	170	Sucrose	30
Na$_2$-EDTA	37.3		
FeSO$_4$7H$_2$O	27.8	Hormone	Concentration (μM)
MnSO$_4$4H$_2$O	22.3	2,4-D	1
ZnSO$_4$7H$_2$O	8.6	6-BA	1
H$_3$BO$_3$	6.2		
KI	0.83		
Na$_2$MoO$_4$2H$_2$O	0.25	pH	5.8–6.0 (before autoclaving)
CoCl$_2$6H$_2$O	0.025		
CuSO$_4$5H$_2$O	0.025		

eighth generation was maintained in a similar medium, without agar, on a rotary shaker at 25°C for 6 weeks (Sugisawa and Ohnishi, 1976).

Cell line selection

Selection experiments have already yielded a large number of mutants. There are three approaches of cell line cloning: by plating cell aggregates, by protoplast culture, and by single cell culture. Selection of high-producing cell line by the method of cell aggregate cloning is as follows (Yamamoto *et al.*, 1982). The calli were cut into many segments (volume, *ca.* 3 mm^3) with a scalpel. Each segment was coded and placed on agar-medium (25 ml) in a sectioned Petri dish 9 cm in diameter. The agar-medium consisted of the liquid medium and 0.8 % (w/v) agar. The segments were cultured under suitable conditions for a certain period. Each of the 9 segments on a Petri dish was cut into two cell-aggregates; one (D$_1$) for subculture and the other (D$_2$) for quantitative analysis of the pigment. From the analysis of D$_2$, we selected the reddest D$_1$ cell-aggregate from each Petri dish. These selected cell-aggregates were cut into several segments (volume, *ca.* 3 mm^3). All these segments were coded and transplanted onto fresh medium in a 9-section Petri dish. This selection procedure was repeated many times.

Suspension Culture

Flask culture

Flask suspension cultures were obtained by the transfer of friable callus lumps to an agitated liquid medium of the same composition as that used for the growth of callus. A relatively large initial inoculum was advantageous, as this would ensure that sufficient single cells and/or small clumps were released into the medium to provide a suitably

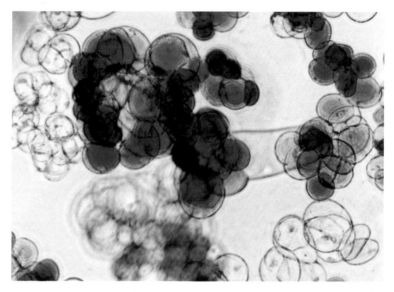

Figure 1   Photograph of suspension cells of *Perilla frutescens* cultured in a shake flask

high cell density for subsequent growth. Rotational speeds of the orbital shakers should be in the range of 30–150 rpm with an orbital motion stroke of 2–4 cm.

For example, suspension cells of *P. frutescens* (Figure 1), which produced a high level of anthocyanin, were cultured in a 500 ml conical flask containing 100 ml LS medium with the addition of 3 % sucrose, 1 µM 2,4-D and 1 µM benzylamino-purine (BA). The cells were incubated on a rotary shaker (75 rpm) at 25°C under continuous light irradiation supplied by ordinary fluorescent lamps (17–20.4 W/m²). The subculture period and inoculum density were generally controlled at 7–10 days and 25 g wet cells/L, respectively (Zhong *et al.*, 1991).

Bioreactor culture

Large-scale culture of plant cells has developed from the need to study the problems of scale-up in the development of commercial processes for the production of biomass or secondary products. However, the growth of plant cell suspensions in bioreactors also allows the study of the effects of conditions such as aeration, oxygen and carbon dioxide levels on growth and secondary product formation, a study not possible in shake flasks.

For bioreactor operation, a reactor was filled with a certain amount of medium and autoclaved at 121°C for 30 min prior to the start of a cultivation. After the reactor had cooled to room temperature, the agitation speed and aeration rate were set as required. The cultivation temperature was set and then automatically controlled as required. Inoculation was performed by pouring the inoculum through a large opening in the head plate. The inoculum density was the same as in flask cultivation.

Table 3   Pigment content of *P. frutescens* after cell line selection

Selection generation	3rd	4th	5th
Pigment content (mg/g dry cell)	80.6	196.4	197.1

FACTORS AFFECTING METABOLITE FORMATION BY *PERILLA* CELLS

Cell Line, Cell Aggregate Size, Subculture Period and Inoculum Density

Cell line

In the production of anthocyanins (pigment) and rosmarinic acid (phenylpropanoid) from cultured callus tissue of Akachirimen-shiso (*Perilla* sp.), Tamura *et al.* (1989) claimed that the relative amount of certain anthocyanins produced by the callus tissue was different and greater than found in the intact plant. However, by means of several subcultures, in which the pigmented cell line was selected, the anthocyanins of the callus were changed in amount and nature to that found in the intact plant. This change might be attributable to gene mutation in the cultured cell.

In anthocyanin production by *P. frutescens* cells, a high-pigment-producing strain in the cultured cells was selected through the method of cell-aggregate cloning as described above. The result is shown in Table 3 (Yoshida, M., Master thesis, Osaka Univ., 1989), in which the pigment content was measured as described elsewhere (Zhong *et al.*, 1991).

Cell aggregate size

The influence of the size of the cell-aggregate in suspension culture of *P. frutescens* on anthocyanin accumulation was investigated by inoculating with cell-aggregates of a screened size in successive subcultures, while for the control, the cell aggregate sizes were without screening. As shown in Table 4, compared with the control, the anthocyanin

Table 4   Effect of sizes of cell aggregates on anthocyanin accumulation by *P. frutescens* cells subcultured successively in a shake flask[1]

Size (μm)	Anthocyanin content (mg/g dry cell) Subculture				
	1st	2nd	3rd	4th	5th
Control	90	87.4	81.7	90.2	77.9
250–2000	67.6	56.2	66.2	54.5	42.1
149–250	76.3	50.3	81.8	79.7	81.4
37–149	105	71.9	88.7	85.5	64.9

[1]The cell growth was almost the same in all the above cases. The data shown here was an average of at least 3 samples. The cell aggregate sizes of the control were in the range of 37–2000 μm with a distribution similar to that as described (Zhong *et al.*, 1992).

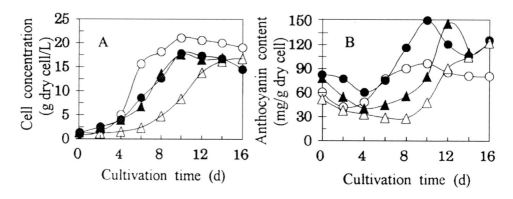

Figure 2 Effect of subculture period on anthocyanin accumulation by *P. frutescens* cells in a flask culture. Symbols for subculture period: Open circle, 5 d; closed circle, 7 d; closed triangle, 10 d; open triangle, 14 d

content of flask cultures which were inoculated with cell aggregate sizes greater than 250 μm was lower. The pigment content in other cultures, which were inoculated with cell aggregate sizes smaller than 250 μm, was almost the same as that of the control. The result suggests that cell-to-cell communication, which may be important to the metabolite formation, probably depends on a certain range of cell aggregate sizes.

Subculture period

The effect of subculture periods on *P. frutescens* cell cultures was studied by subculturing the cells at an interval of 5, 7, 10 or 14 days (Zhong and Yoshida, 1994d). The results indicated that the cells subcultured at an interval of 7 or 10 days yielded a higher anthocyanin content, i.e. higher specific anthocyanin production (mg/g dry cells), compared with those at an interval of 5 or 14 days. In these cases, the anthocyanin content proved to be unstable during subcultures. The results indicated that alteration of the subculture period is one way of overcoming the instability of metabolite production in plant cell cultures, which is one of the main obstacles to the commercialisation of plant cell culture processes.

A further investigation on the growth and production dynamics of the cell cultures at different subculture periods indicated that although the cells which were subcultured at an interval of 5 days propagated relatively faster, the anthocyanin content was rather lower, compared with those at an interval of 7 and 10 days. The cells subcultured at a 14-day interval showed lower cell growth and anthocyanin content, compared with those at a 7- or 10-day interval (Figure 2). The results indicate that the physiological and metabolic aspects of cultured cells varied greatly with different subculture periods.

Inoculum density

The effect of inoculum density (15, 25, 50 and 75 g wet cells/L) on cell growth and anthocyanin formation by suspension cells of *P. frutescens* was investigated by using a 500 ml conical flask containing 100 ml medium (Zhong and Yoshida, 1995). The results showed that the growth rate and cell concentration reached at around 10th day were quite similar for different inoculum densities, and the substrate (sugar) was also almost completely consumed by that day in all the cases (Zhong and Yoshida, 1995). However, the anthocyanin content (i.e. specific anthocyanin production, mg/g dry cells) and total anthocyanin production (g/L) were very different at various inoculum densities. The production and specific production of the pigment reached the highest at an inoculum density of 50 g wet cells/L (Zhong and Yoshida, 1995).

Chemicals

Carbon source

In plant cell cultures, sucrose is usually a suitable carbon source. In *P. frutescens* cultures, Harada (Harada, H., Master thesis, Osaka Univ., 1988) and Koda *et al.* (1992) found that among different carbon sources, sucrose was found to be the best one for both the cell growth and formation of anthocyanin pigment.

In high density batch cultures of *P. frutescens* (at an inoculum size of 50 g wet cells/L), the initial sucrose concentration showed a conspicuous effect on the kinetics of cell growth, sugar consumption, and anthocyanin production by *P. frutescens* cells. The maximum cell density of 38.3 g dry cells/L was obtained after 11 days' cultivation at an initial sucrose concentration of 60 g/L, while the highest pigment production of more than 5.8 g/L was attained at 45 g/L of sucrose (Zhong and Yoshida, 1995).

In addition, the data indicated that the initial sucrose concentration affected the excretion of anthocyanin by cell cultures of *P. frutescens* (Zhong *et al.*, 1994c). When the medium contained 40–50 g/L of sucrose a much higher amount of anthocyanin was released from the cultured cells, compared with the control of 30 g/L of sucrose. Further experiments confirmed this result.

Nitrogen source

An investigation of the effect of the nitrogen source on the growth and anthocyanin production by *P. frutescens* cells has been carried out. The results indicated that the total amount of nitrogen and the ratio of nitrate to ammonium salts in the LS medium were the most suitable for the cell cultures (Harada, H., Master thesis, Osaka Univ., 1988). But a study made by another group showed that a NO_3^-/NH_4^+ of 10 with a total nitrogen of 30 mM gave the best results for the growth of, and pigment formation by, *P. frutescens* cultures (Koda *et al.*, 1992).

Plant growth regulator

Table 5 shows that caffeic acid of 1.1–2.7 mg per g fresh weight of calluses were estimated to be present in *Perilla* calli. The yield from the cells in MS-III medium containing

Table 5 Caffeic acid formation in *Perilla* callus cultured in various media

Agar media	Growth regulators (mg/L)			Fresh wt. (g) of callus subcultured	Fresh wt. (g) of callus cultured for 21 days	Caffeic acid (mg) per g fresh wt. of the callus
	2,4-D	NAA	kinetin			
SH-M	0.5	—	0.01	3.89	18.32	1.82
MS-I	1.0	—	0.1	3.64	20.00	1.53
MS-II	1.0	—	5.0	4.15	20.18	2.67
MS-III	—	1.0	0.1	4.01	12.11	2.60
G	1.0	—	—	3.85	19.45	1.12

1-naphthyl acetic acid (NAA, 1.0 mg/L) instead of 2,4-D and in MS-II medium containing a high concentration of kinetin (5.0 mg/L) were about double those in MS-I and G media (Ishikura *et al.*, 1983).

Shin (1985) reported that an addition of one ppm of NAA instead of 2,4-D in the modified MS medium containing 5 ppm of kinetin caused 75% increase in the growth of *P. frutescens* cells and two-fold increase in production of essential oils.

In anthocyanin production, Harada's result indicated that 1 μM of 2,4-D in combination with 1~10 μM of BA in LS medium was the most suitable for cell growth and pigment formation (Harada, H., Master thesis, Osaka Univ., 1988) whereas Koda *et al.* (1992) claimed that 10 μM NAA in combination with 1 μm BA was the best for *P. frutescens* cultures (Table 6).

Precursor

In a two stage culture of the young leaf of *Perilla* species, Shin (1986) reported that the addition of mevalonic acid lactone (10 ppm) increased the wet cells and yield of essential oil (including sesquiterpene hydrocarbons and sesquiterpene alcohol) of callus from 1.8 to 2.6 g and from 4.7 to 18.7 mg/tube, respectively.

Table 6 Effects of phytohormones on growth and pigment production by *P. frutescens* cells

Auxin	Cytokinin	Fresh wt. (g/flask)	Pigment	
			(CV/g)	(CV/flask)
2,4-D 10^{-5} M	BA 10^{-6} M	3.80	0.66	2.51
IAA		1.09	1.54	1.68
NOP		3.21	4.25	13.64
NAA 10^{-4} M		2.81	3.87	10.87
NAA 10^{-5} M	BA 10^{-5} M	3.52	6.12	21.54
	BA 10^{-6} M	3.21	6.84	21.96
	BA 10^{-7} M	1.85	4.84	8.95
NAA 10^{-6} M	BA 10^{-6} M	1.99	3.15	6.27

CV = [OD$_{524}$ (Sample wt. (g) + 10)]/Sample wt. (g).
Abbreviations: BA, benzylamino-purine; 2,4-D, 2,4-dichlorophenoxyacetic acid; IAA, indole-3-acetic acid; NAA, 1-naphthaleneacetic acid; NOP, naphthoxypropionic acid.

Table 7   Growth and volatile matter content in *Perilla* callus fed with and without mevalonate

	Without	*With mevalonate*			
		Non-labeled	*2,2-²H₂-*	*4,4-²H₂*	*5,5-²H₂*
Growth index	5.87	1.29	1.29	1.29	1.06
Volatile content (10⁻³% fresh wt.)	0.067	1.83	—	—	—
Sesquiterpene content (10⁻³% fresh wt.)	0.013	0.025	—	—	− +
Sesquiterpenes					
Isolongifolene	+		+		+
α-Curcumene	+	+		+	
β-Bisabolene	+	+	+	+	+
Chamigrene	+		+		+
Thujopsene	+	+			+
Cuparene	+	+	+	+	+
α-Farnesene	+	+			
β-Farnesene	+	+			
Farnesol	+	+			

-, not determined; +, detected by GC/MS analysis.

Later, Nabeta *et al.* (1993) claimed that the administration of mevalonate (MVA, 10 mmol/8 ml culture medium) significantly retarded the growth of *Perilla* callus tissues [growth index (fresh wt.): 1.06~1.29 with MVA and 5.87 without, see Table 7]. However, a significant increase in the level of total volatiles was observed (24 times higher than that in the calli grown without MVA, see Table 7). The increase in the total volatile content was mainly due to the formation of long-chain compounds including palmitic acid, tetradecane, 2,4-decadienal and butyric acid. The level of sesquiterpene hydrocarbons in the calli with MVA was 1.8 times higher than that in the calli without. Sesquiterpene hydrocarbons were always observed in the calli cultured with the deuterated MVAs, and deviations in their qualitative patterns were observed upon the addition of the differently deuterated MVAs. Cuparene and β-bisabolene, however, were always observed.

Inhibitor

In order to clarify the biosynthetic regulation of caffeic acid in *Perilla* cell suspension cultures, Ishikura *et al.* (1983) examined the response of the cells to three inhibitors, L-2-aminooxy-3-phenylpropionic acid (L-AOPP), 2-aminooxyacetic acid (AOA) and N-(phosphonomethyl) glycine (glyphosate). The administration of L-AOPP, AOA and glyphosate to the *Perilla* cells inhibited caffeic acid formation to a large extent. An 80% inhibition of caffeic acid formation was caused by 10^{-4} M L-AOPP whereas phenylalanine and tyrosine contents of the cells became 7.5 and 2.3 times higher at this L-AOPP concentration than those in the control. An 85% inhibition of caffeic acid formation was achieved at 10^{-3} M glyphosate concentration, while 10^{-3} M AOA inhibited caffeic acid formation by 95% and also growth rate by 80%.

Table 8   Comparison of parameter values in cultivation of *P. frutescens* cells at different temperatures in flask cultures

T (°C)	μ (d⁻¹)	X_{max} (g/L)	AC_{max} (mg/g)	TA (g/L)	$Y_{X/S}$ (g/g)	$Y_{P/S}$ (g/g)	Q (mg/L/d)
22	0.21	21.6	185.1	3.67	0.70	0.115	211
25	0.32	19.9	176.9	3.52	0.66	0.112	268
28	0.37	19.2	67.6	1.25	0.62	0.032	68

T, temperature; μ, specific growth rate; X_{max}, maximum cell concentration; AC_{max}, maximum anthocyanin content; TA, total anthocyanin; $Y_{X/S}$, cell yield; $Y_{P/S}$, anthocyanin yield; Q, volumetric anthocyanin productivity.

In a related work, Ishikura and Takeshima (1984) reported that when 1 mM glyphosate was added to the cell culture in the logarithmic and stationary phases, the amount of caffeic acid ceased to increase and remained at a nearly constant level during the following several days. The shikimic acid content of cells from 14-day culture grown in the presence of 1 mM glyphosate increased up to 74.9 μg per g fresh weight during 6-day culture, whereas that of the control cells was undetectable. The dosage of 0.15 mM L-AOPP, an inhibitor of phenylalanine ammonia-lyase, to the cells did not cause shikimic acid accumulation.

Temperature

The effect of culture temperature on cell growth and anthocyanin formation by the cell cultures of *P. frutescens* has been investigated (Zhong and Yoshida, 1993b). The results showed that at different incubation temperatures (i.e. 22°C, 25°C and 28°C), although the maximum cell concentration obtained was identical (about 20 g/L), the growth rate, anthocyanin content and total anthocyanin produced were very different. At a higher temperature, in a range of 22~28°C, a higher specific growth rate was obtained. However, anthocyanin production, its productivity and yield were remarkably reduced at 28°C compared with those at 22°C and 25°C, respectively (Table 8). The highest anthocyanin productivity was obtained at 25°C. The above results indicated that the culture temperature was an important factor regulating the mechanism of anthocyanin biosynthesis in cultured cells of *P. frutescens*, and that this parameter should be strictly controlled during these cultivations.

Light Irradiation

In plant cell-tissue cultures, light has stimulatory, inhibitory, or insignificant effects on cell growth and the accumulation of plant metabolites. For example, in modified MS medium containing 1 ppm of 2,4-D and 5 ppm of kinetin, Shin (1985) claimed that light (irradiation at 1600 lux) decreased the callus growth by 25% but rather increased the content of essential oil by two-fold, i.e. from 2% to 4%.

Table 9 Effect of light source on cell growth and anthocyanin content of *P. frutescens* cells cultivated for 12 days in 500 ml conical flasks

Light spectrum	Growth (g wet cells/100 mL)	Anthocyanin content (mg/g dry cells)
Ordinary fluorescent lamp	27.3	296.1
Plant lamp	29.8	239.7
UV lamp	16.6	187.2

*The ordinary fluorescent lamp has light spectrum range of 450~610 nm; the plant lamp possesses two peaks in its light spectrum at 460 nm (blue) and 655 nm (red) for each; the UV lamp has a light spectrum peak at 325 nm. Light irradiation intensity was 1000~1100 lux in all the cases.

Until now, detailed investigations on the optimisation of light irradiation conditions are very scarce. More seriously, there is a lack of information concerning the optimisation of light irradiation on a bioreactor scale, although such an investigation is essential to the design and optimisation of large scale processes for metabolite production by plant cell cultures. Here, a systematic study on the effect of light irradiation on anthocyanin production by *P. frutescens* cells is described.

Influence of light spectrum on anthocyanin accumulation

We found (Table 9) that irradiation from the light source of an ordinary fluorescent lamp with a spectrum range of 450~610 nm was most effective for anthocyanin production by *Perilla* cells, while irradiation by a plant lamp decreased the anthocyanin content a little, and irradiation by a UV lamp showed inhibition to both growth and pigment formation of the cultured cells. In other studies, Ota (1986) claimed that perilla pigment was produced by exposing to a 450–500 nm bluish fluorescent lamp, while no pigment production occurred with a 580 nm daylight fluorescent lamp.

Effects of irradiation period and light intensity on anthocyanin formation in a flask or roux bottle

The effect of light irradiation period on anthocyanin content of *P. frutescens* cells was investigated (Zhong *et al.*, 1991). Although the cell mass was increased 12 fold compared with the inoculum in all the cases, the anthocyanin content of the cells was very different for each case. Light irradiation for the first 7 days or the whole cultivation period was effective for anthocyanin production, while only a small amount of anthocyanin was obtained in the cases of no light irradiation in the first 7 days, even when light was supplied in the subsequent 7 days.

Investigation of the effect of light intensity on pigment production was attempted by using a 500 ml roux bottle with a lighting area of 164 cm^2 at 0.4 vvm (Zhong *et al.*, 1991). The results demonstrated that after 14 days' cultivation the anthocyanin content and total production increased with the increase of light intensity up to 27.2 W/m^2 of the daylight irradiation supplied by an ordinary fluorescent lamp.

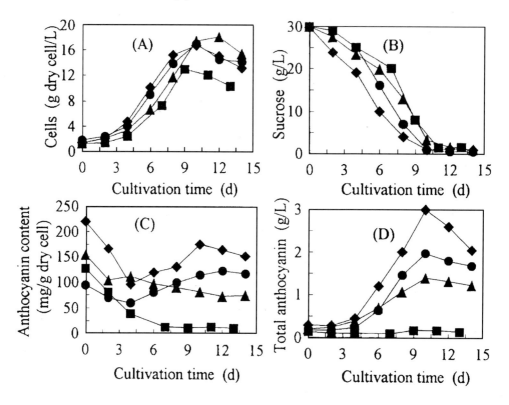

Figure 3 Effect of light irradiation on anthocyanin production in cell cultures of *P. frutescens* in a 2-L (working volume) bubble column reactor. Symbols for light irradiation intensity (W/cm^2): triangle, 0; circle, 13.6; rhombus, 27.2; square, 54.4

Effect of light irradiation on anthocyanin production in a bioreactor

From the above results, to translate the small scale result to a bioreactor scale, it was clearly necessary to investigate what light irradiation intensity was most suitable for anthocyanin production by cultured cells of *P. frutescens* on the bioreactor scale. Experiments in a 2-L (working volume) bubble column reactor were made with light irradiation at different intensities while maintaining the other cultivation conditions the same. Figure 3 indicates that in the bioreactor cultivation without light irradiation, the total anthocyanin was 1.4 g/L. In a cultivation performed at 13.6 W/m^2 of light irradiation intensity, on the 10th day of the cultivation about 2 g/L of total anthocyanin was accumulated. When the light irradiation intensity was increased up to 27.2 W/m^2, a maximum anthocyanin of 3.0 g/L was obtained on the 10th day of the cultivation and this value is comparable to that accumulated in the flask culture where 3.5 g/L of total anthocyanin was formed after 12 days incubation (Zhong *et al.*, 1991). The pigment productivity (g/L/day) reached 2.5 and 1.5 fold compared with that without lighting or with lighting at 13.6 W/m², respectively (Figure 3), and it was even a little higher than that in a flask culture. With a further increase of lighting intensity to 54.4 W/m², however, the total anthocyanin was markedly reduced (Figure 3). This means that such a high irradiation intensity was harmful to pigment formation. The results obtained above may serve as a guide to the optimisation of light irradiation in the scale up of the cultivation process for mass production of the anthocyanin.

Oxygen Supply

Figure 4A shows a typical example of the time course of a cultivation in a 2-L turbine reactor (at 150 rpm and 0.2 vvm). The maximum amount of anthocyanin produced was 0.48 g/L on day 11. The highest cell concentration, about 17 g dry cell/L, was also reached on that day. The anthocyanin content continuously decreased from 65 mg/g dry cell at the beginning to 25 mg/g dry cell at the end of cultivation.

The reasons why anthocyanin yield was poor under the above conditions had to be elucidated. Because the initial volumetric oxygen transfer coefficient ($k_L a$) value was low (ca. 6.8 h⁻¹) (Zhong et al., 1993a), it was considered possible to enhance pigment formation by improving oxygen supply to obtain results like those in cultivations using a shake flask (Zhong et al., 1993a). Experiments were made by using a modified reactor with a sintered sparger to observe the effect of oxygen supply on anthocyanin production by the cells. In the modified reactor, at 150 rpm initial $k_L a$ values of 9.9 and 15.2 h⁻¹ were obtained at 0.1 and 0.2 vvm, respectively (Zhong et al., 1993a).

Figure 4B shows that the total amount of anthocyanin produced at aeration rates of 0.1 and 0.2 vvm was 0.55 g/L on day 10 and 1.65 g/L on day 12, respectively, while cell

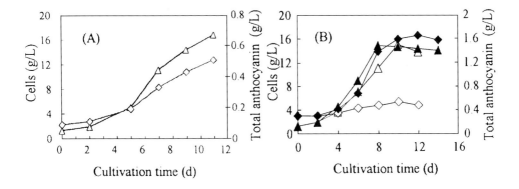

Figure 4 Effect of oxygen supply on anthocyanin production in cell cultures of *P. frutescens* in a 2-L (working volume) stirred bioreactor at 150 rpm. A, ring sparger (at 0.2 vvm); B, sintered sparger (at 0.1 and 0.2 vvm). Symbols: triangle, cell concentration; rhombus, anthocyanin production. In Figure 4B, white and dark symbols are for an aeration rate of 0.1 and 0.2 vvm, respectively.

growth was virtually the same for both cases. Anthocyanin content decreased continuously from 250 mg/g dry cell to 35 mg/g dry cell on day 10 at a 0.1 vvm aeration rate. At an aeration rate of 0.2 vvm, the pigment content decreased from 220 mg/g dry cell to 75 mg/g dry cell on day 6, but after that increased to 120 mg/g dry cell on day 12. At an aeration rate of 0.1 vvm using the sintered sparger, the production of anthocyanin was poor, showing almost the same result as that at 0.2 vvm using a ring sparger (Figure 4A), when the other cultivation conditions were maintained the same. However, when the aeration rate was increased to 0.2 vvm with the new sparger being used, product accumulation increased almost 3-fold, i.e. 1.65 g/L was obtained compared to that at 0.1 vvm using the same sintered sparger or at 0.2 vvm with the ring sparger. The results indicate that the oxygen supply conditions affected product biosynthesis.

In addition, regardless of the different cultivation conditions mentioned above, we observed a similar pattern of change in dissolved oxygen (DO) level during cultivation, with the lowest DO value being around 10–20% of air saturation. An investigation of the effect of DO level on anthocyanin production indicated that there was no difference in production at a DO level controlled at 20% or 80% of saturation (unpublished results). It is suggested that the phenomenon observed was due to differences in the oxygen uptake rate (OUR) between cells producing anthocyanin and those not producing anthocyanin, with the OUR of the former being larger. We confirmed such a difference in OUR between more anthocyanin-producing cells and less anthocyanin-producing cells (Zhong et al., 1994b).

Shear Stress

Plant cells are usually sensitive to hydrodynamic stress as each usually has a large volume (ranging from 20 to as much as 100 µm in diameter) and a rigid cellulose-based inflexible cell wall, and often has a very large vacuole, comprising up to 95 % or more of the cell volume. Such characteristics of plant cells imply that they are easily susceptible to damage under a certain degree of shear. In our case, the detrimental effects of shear stress on *P. frutescens* cells were demonstrated in both short-term experiments and batch cultivations in bioreactors (Zhong et al., 1993a and 1994a).

The quantitative shear effects on cell growth and anthocyanin production by the cell culture were analysed in a series of batch cultures in a 3-L (working volume) plant cell reactor with a marine impeller having a diameter of 85 mm (larger impeller) or 65 mm (smaller impeller). Figure 5 shows the effects of the average and maximum shear rates on the specific growth rate and specific production of anthocyanin (i.e. anthocyanin content) of *P. frutescens* cells in the bioreactor cultivations. Here, the maximum shear rate is assumed to be represented by the impeller tip speed (ITS). The specific growth rate was apparently reduced at an average shear rate (ASR) over 30 s^{-1} or at an ITS of over 8 dm/s. The anthocyanin content was relatively high at an ASR below 30 s^{-1} or an ITS below 8 dm/s. At higher shear rates, the pigment content of the cultured cells showed an obvious decrease. The effects of shear on the maximum cell concentration, total anthocyanin production and the volumetric productivity of anthocyanin were similar to those as shown in this figure. Relatively high cell concentration, pigment production and pigment productivity, as well as cell and pigment yields, were achieved in the bioreactor

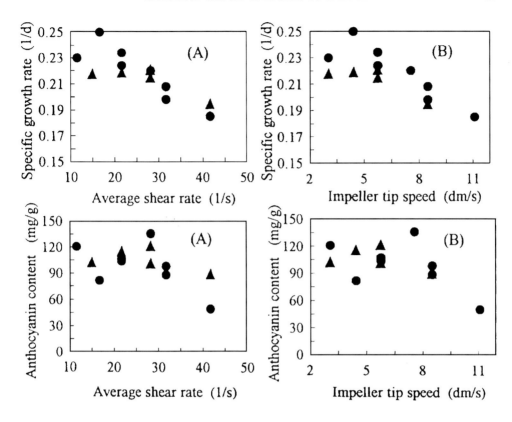

Figure 5   Effect of shear on the specific growth rate and anthocyanin content in cell cultures of *P. frutescens* in a 3-L (working volume) plant cell bioreactor with a marine impeller. A, average shear rate; B, impeller tip speed. Symbols: circle, big impeller; triangle, small impeller

at an ASR of 20~30 s^{-1} or at an ITS of 5~8 dm/s (Zhong *et al.*, 1994a). These criteria were equivalent to the following operational conditions in the reactor: an agitation speed of 120~170 rpm using the larger impeller (85 mm in diameter). At an ASR of over 30 s^{-1} or an ITS of over 8 dm/s, the cell growth, pigment production and its specific production, as well as the yields of the cells and the pigments were poor due to the detrimental effects of the shear stress on the cultured cells.

CONCLUSIONS AND INDUSTRIAL PERSPECTIVES

In this chapter, the basic techniques for cell and tissue cultures of *Perilla*, including callus induction, cell line selection as well as establishment of suspension cultures, were introduced; and the progress in the cell and tissue cultures since the 1970s has been reviewed. The factors affecting the cell and tissue cultures of *Perilla* were cell line, cell aggregate size, inoculum density, culture age (biological factors); light irradiation, oxygen supply, shear stress, culture temperature (physical factors); precursor, inhibitor, and medium components (including carbon and nitrogen sources as well as plant growth regulators) (chemical factors). A systematic study on the optimisation of these factors for anthocyanin production was shown.

As demonstrated in our previous work (Zhong *et al.*, 1991), the bioprocess of anthocyanin production by suspension cells of *P. frutescens* was successfully scaled up from a shake flask to a bioreactor scale (with working volume of 3 L) by paying attention to the factors of light irradiation, oxygen supply and shear stress. Through a similar approach, the cell cultivation in a 100-L jar fermentor was succeeded by San-ei Chemical Industry Co. (Osaka, Japan) (Koda *et al.*, 1992); the red pigment produced in the big fermentor was 7-fold increase compared with its inoculation level, and the amount was estimated to be equivalent to that extracted from 35 kg of fresh perilla leaves (Koda *et al.*, 1992). For commercial production of anthocyanin pigment by the cell cultures, however, much further improvement in the efficiency of the San-ei process was necessary.

In the case of famous shikonin process developed by Mitsui Petrochemicals Co., the shikonin productivity by cell cultures and the market value of the product in 1987 were 0.1 g/L/d and US$ 4000/kg, respectively (Ilker, 1987). In contrast, at that time the pure anthocyanin was sold at US$ 1250~2000/kg, while the highest pigment productivity obtained via cell culture was only 55 mg/L/d (Yamakawa *et al.*, 1983). Due to the relatively cheaper value and lower yield of anthocyanin compared with those of shikonin, it is apparent that further technological improvements were required for its practical industrial exploitation. In our recent work, a high anthocyanin productivity of ca. 0.3 g/L/d was achieved (in a bioreactor) (Zhong *et al.*, 1991), and more recently it was further enhanced up to 0.58 g/L/d (in a shake flask) (Zhong and Yoshida, 1995). Here, it may be reasonable to consider that a big step has been taken towards the commercial anthocyanin production by cell cultures. Future work includes further bioreactor scale-up as well as the product safety test (especially for its application in food industries).

ACKNOWLEDGEMENTS

The authors' work described was conducted at ICBiotech, Osaka University. During the manuscript preparation, JJZ was supported by Japan Society for the Promotion of Science and the State Education Commision of China.

REFERENCES

Dixon, R.A. (1985) Isolation and maintenance of callus and cell suspension cultures. *Plant cell culture: a practical approach*, ed. R.A. Dixon. IRL Press, Oxford. pp. 1–20.
Furukubo, M., Tabata, M., Terada, T., and Sakurai, M. (1989) Manufacture of chlorphenesin carbamate glycoside by plant tissue culture. *Japan Kokai Tokkyo Koho*, JP01-38095.
Ilker, R. (1987) *In vitro* pigment production: an alternative to color synthesis. *Food Technology*, **41**, 70–72.

Ishikura, N., Iwata, M. and Mitsui, S. (1983) The influence of some inhibitors on the formation of caffeic acid in cultures of *Perilla* cell suspensions. *The Botanical Magazine, Tokyo*, **96**, 111–120.

Ishikura, N., and Takeshima, Y. (1984) Effects of glyphosate on caffeic acid metabolism in *Perilla* cell suspension cultures. *Plant and Cell Physiology*, **25**, 185–189.

Koda, T., Ichi, T., Yoshimitu, M., Nihongi, Y. and Sekiya, J. (1992) Production of *Perilla* pigment in cell cultures of *Perilla frutescens*. *Nippon Shokuhin Kogyo Gakkaishi*, **39**, 839-844 (in Japanese).

Nabeta, K., Oda, T., Fujimura, T., and Sugisawa, H. (1984) Biosynthesis of cuparene from mevalonic acid-6,6,6-^{2}H$_3$ by *in vitro* callus culture of *Perilla* sp. *Agricultural and Biological Chemistry*, **48**, 3141–3143.

Nabeta, K., Oda, T., Fujimura, T. and Sugisawa, H. (1985) Metabolism of RS-mevalonic acid-6,6,6-^{2}H$_3$ by *in vitro* callus culture of *Perilla* sp. *Agricultural and Biological Chemistry*, **49**, 3039–3040.

Nabeta, K., Kawakita, K., Yada, Y. and Okuyama, H. (1993) Biosynthesis of sesquiterpenes from deuterated mevalonates in *Perilla* callus. *Bioscience, Biotechnology, and Biochemistry*, **57**, 792–798.

Ota, S. (Kinjirushi Wasabi K.K.) (1986) Perilla pigment production by callus cultivation. *Japan Kokai Tokkyo Koho*, JP61-195688.

Shin, S.H. (1985) Studies on tissue culture of *Perilla frutescens* var. acuta (I). *Saengyak Hakkoechi*, **16**, 210-213 (in Korean).

Shin, S.H. (1986) Studies on tissue culture of *Perilla* species. *Saengyak Hakkoechi*, **17**, 7–11 (in Korean).

Sugisawa, H. and Ohnishi, Y. (1976) Isolation and identification of monoterpenes from cultured cells of *Perilla* plant. *Agricultural and Biological Chemistry*, **40**, 231–232.

Tabata, M., Umetani, Y., Oya, M. and Tanaka, S. (1988) Glucosylation of phenolic compounds by plant cell cultures. *Phytochemistry*, **27**, 809–813.

Tamura, H., Fujiwara, M., and Sugisawa, H. (1989) Production of phenyl propanoids from cultured callus tissue of the leaves of Akachirimen-shiso (*Perilla* sp.). *Agricultural and Biological Chemistry*, **53**, 1971–1973.

Tanimoto, S. and Harada, H. (1980) Hormonal control of morphogenesis in leaf explants of *Perilla frutescens* Britton var. crispa Decaisne f. viridi crispa Makino. *Annals of Botany (London)*, **45**, 321–327.

Terada, T., Yokoyama, T. and Sakurai, M. (1989) Optical resolution of propranolol hydrochloride or pindolol by cell culture of *Perilla frutescens* var. cripsa or *Gardenia jasminoices*. *Japan Kokai Tokyo Koho*, JP01 225498.

Tomita, Y. and Ikeshiro, Y. (1994) Biosynthesis of ursolic acid in cell cultures of *Perilla frutescens*. *Phytochemistry*, **35**, 121–123.

Yamakawa, T., Kato, S., Ishida, K., Kodama, T. and Minoda, Y. (1983) Production of anthocyanins by *Vitis* cells in suspension cultures. *Agricultural and Biological Chemistry*, **47**, 2185–2191.

Yamamoto, Y., Mizuguchi, R., and Yamada, Y. (1982) Selection of a high and stable pigment-producing strain in cultured *Euphorbia millii* cells. *Theoretical and Applied Genetics*, **61**, 113–116.

Zhong, J.-J., Fujiyama, K., Seki, T. and Yoshida, T. (1994a) A quantitative analysis of shear effects on cell suspension and cell culture of *Perilla frutescens* in bioreactors. *Biotechnology and Bioengineering*, **44**, 649–654.

Zhong, J.-J., Konstantinov, K.B. and Yoshida, T. (1994b) Computer-aided on-line monitoring of physiological variables in suspended cell cultures of *Perilla frutescens* in a bioreactor. *Journal of Fermentation and Bioengineering*, **77**, 445–447.

Zhong, J.-J., Seki, T., Kinoshita, S. and Yoshida, T. (1991) Effect of light irradiation on anthocyanin production by suspended culture of *Perilla frutescens*. *Biotechnology and Bioengineering*, **38**, 653–658.

Zhong, J.-J., Seki, T., Kinoshita, S. and Yoshida, T. (1992) Rheological characteristics of cell suspension and cell culture of *Perilla frutescens*. *Biotechnology and Bioengineering*, **40**, 1256–1262.

Zhong, J.-J., Xu, G.-R. and Yoshida, T. (1994c) Effects of initial sucrose concentration on excretion of anthocyanin pigments in suspended cultures of *Perilla frutescens* cells. *World Journal of Microbiology and Biotechnology*, **10**, 590–592.

Zhong, J.-J., Yoshida, M., Fujiyama, K., Seki, T. and Yoshida, T. (1993a) Enhancement of anthocyanin production by *Perilla frutescens* cells in stirred bioreactor with internal light irradiation. *Journal of Fermentation and Bioengineering*, **75**, 299–303.

Zhong, J.-J. and Yoshida, T. (1993b) Effects of temperature on cell growth and anthocyanin production by suspension cultures of *Perilla frutescens* cells. *Journal of Fermentation and Bioengineering*, **76**, 530–531.

Zhong, J.-J. and Yoshida, T. (1994d) Effects of biological factors and initial sucrose concentration on anthocyanin production by suspended cultures of *Perilla frutescens*. *Better Living Through Innovative Biochemical Engineering*, eds. WK Teo, MGS Yap, and SKW Oh. National University of Singapore, Singapore. pp. 155–157.

Zhong, J.-J. and Yoshida, T. (1995) High density cultivation of *Perilla frutescens* cell suspensions for anthocyanin production: effects of sucrose concentration and inoculum size. *Enzyme and Microbial Technology*, **17**, 1073–1079.

4. APPLICATIONS AND PRESCRIPTIONS OF PERILLA IN TRADITIONAL CHINESE MEDICINE

YUH-PAN CHEN

Brion Research Institute of Taiwan, 116, Chungching S. Rd., Sec. 3, Taipei, Taiwan 100

INTRODUCTION

Since the advent of "Shen Nong Ben Cao Jing" (Shen Nong's Herbal), the progenitor of herbals in traditional Chinese medicine, completed around 25 A.D., which classified herbal drugs into upper grade, mid-grade and lower grade, all subsequent herbals classified Chinese herbal drugs according to this tradition. The upper grade drugs are known as the imperial drugs which are non-poisonous and are used mainly for nurturing our lives; the mid-grade drugs are known as the ministerial drugs which are either non-poisonous or poisonous and are used chiefly to nurture our temperament; and the lower grade drugs are known as the assistant or servant drugs which are used for treating disease and are mostly poisonous.

In clinical diagnosis, a physician of traditional Chinese medicine will first consider the circulation of *qi*, blood and water. The so-called blood conformation in traditional Chinese medicine (a conformation in traditional Chinese medicine can be approximated to a symptom complex or syndrome in Western medicine) refers to "blood stasis" which is a poor blood circulation condition resulted from congestion or stagnation of blood in the body and may lead to formation of disease. A water conformation is also referred to as "water-stagnancy conformation" which designates poor water metabolism whereof unbalanced circulation and distribution of water may cause disease. Whereas, *qi* is also an essential element of life and is often the element pertinent to living, senescence, disease and death. *Qi* is invisible but mobile, and the pathological state resulted from *qi* impediment or stagnation is known as "*qi* impediment conformation", and a free circulation of *qi* is essential for maintenance of good health.

PERILLA AS RECORDED IN ANCIENT CHINESE MEDICAL CLASSICS

Perilla is recorded in "Ming Yi Bie Lu" (Renown Physicians' Extra Records), a Chinese medical classic completed around 500 A.D., wherein the herb is listed as a mid-grade drug under the name "su". The herb has also been recorded in many other medical works under different names such as "zi su" in "Shi Liao Ben Cao" (Bromatotherapy Herbs), or "chi su" in "Zhou Hou Fang" (Prescriptions Ready at Hand's Reach). Li Shi-zhen (1518–1593 A.D., Ming dynasty) noted in his great work "Ben Cao Gang Mu" (Categorized and Itemized Herbal):

The character "su" means comforting, which implies that the herb comforts our bodies and promotes the circulation of blood and *qi*. Tao Hong-jing (462–536 A.D.,

Liang dynasty) stated, "The herb su is purple coloured on the undersides of its leaves and possesses a very aromatic flavour. The other species that are not purple coloured and do not have an aroma, which resemble *ren* [*Perilla frutescens* (L.) Britton var. japonica (Hassk.) Hara] are known as wild Perilla and have no use as medicine". The "Ming Yi Bie Lu" records this herb, saying, "Perilla is used chiefly to descend *qi*, and remove cold from the central torso of the body; the seed of the herb is especially good for these effects". However, Su Song said, "The "su" species that are purple coloured on the backsides of leaves are better, which are gathered in the summer for the leaves and stems and in the autumn for the seeds, there being several varieties of the herb".

From the above statements we can see that in ancient times, it was the seed of Perilla that was commonly used, and in the Song dynasty (960–1279 A.D.), the stem, leaf and seed of the herb seemed to have been used equally commonly. Nowadays, on the herb market we have supply of this herb in two forms: one is the leaves admixed with stems and the other is simply the seed.

As recorded in the "Ben Cao Gang Mu", the stem and leaf of Perilla are described as having a pungent taste, a warming nature and no toxicity, and are said to be chiefly indicated for relaxing muscles, perspiring skin, dispersing winds and chills, moving *qi*, relieving stomach, resolving phlegm, venting the lungs, harmonising blood, warming stomach, stopping pain, arresting asthma, stabilising embryo, detoxifying poisoning from eating fish and crab and treating snake and dog bites. Perilla seed is said to have similar effects. As for the differences of uses between Perilla leaf and seed, Li Shi-zhen said that both the Perilla leaf and seed were of similar effects, but the leaf was good for dispersing winds and the seed was good for clearing and dredging the upper and lower torsos.

PERILLA AS USED IN CHINESE HERBAL MEDICINE

Among modern literature, it is generally believed that the drug items "Perilla leaf", "Perilla stalk" and "Perilla seed" as used in traditional Chinese medicine are the dried leaves, dried stems, and dried mature seeds of the plant *Perilla frutescens* (L.) Britton var. *crispa* (Thunb.) Decne. of the Labiatae family and plants of related genera (Hsu *et al.*, 1986; Namba, 1980). The Chinese Pharmacopoeia (1990) also includes the dried mature fruits, dried leaves and dried stems of *Perilla frutescens* (L.) Britt. for the articles of Fructus Perillae, Folium Perillae, and Caulis Perillae. The Japanese Pharmacopoeia (1991), on the other hand, lists Herba Perillae as being derived from the leaves and twig ends of *Perilla frutescens* Britton var. *acuta* Kudo and related plants.

According to Chinese medical literature, Perilla possesses a pungent flavour and a warming property, and enters its effects into the lung and the spleen meridians. The Perilla leaf has been described as having the effects of diaphoresis, antipyresis, moving *qi*, relieving central torso (adjusting gastrointestinal functions and aiding digestion), antidoting poisoning from eating fish and crab. Hence it is suitable for treating common cold by rendering diaphoresis to resolve fever, allaying cough and asthma, effecting tranquility, relieving epigastric and abdominal distension, and strengthening the stomach. It is usually used together with other herbs for these effects. Besides, Perilla is also effective against poisoning from eating fish and crab. Whereas, Perilla stalk is effective in soothing *qi* and stabilising the embryo, and is thus capable of treating *qi* adversity, abdomen ache, and embryonic aching and instability.

Perilla seed possesses the effects of descending *qi*, arresting asthma, stopping cough, dissipating phlegm, relieving the chest and resolving depression, wherefore it is indicated for adverse cough, phlegm-associated asthma, *qi* impediment, and constipation. In other words, Perilla leaf is good at dispersing the ailing evils, and Perilla stem is better in regulating *qi*, while Perilla seed is suitable for descending phlegm. Hence, Perilla leaf is mostly used in common cold with chills and fever; Perilla stalk is mostly used in pectoral distress and vomiting and in calming the embryo; and Perilla seed is frequently prescribed for asthmatic cough and for treating phlegm problem (Hsu *et al.*, 1986; Namba, 1980; Takagi *et al.*, 1982).

CLINICAL APPLICATIONS OF PERILLA

Clinically, Perilla is often used in combination with other herbs in treating external contractions of wind and cold evils manifesting chest depression together with nausea, vomiting and other symptoms of the gastrointestinal type of cold. The diaphoretic effect of Perilla leaf is weaker than that of ephedra and cinnamon. Thus, when used alone, it usually does not produce noticeable effect, and therefore has to be used in combination with other herbs such as schizonepeta, siler, fresh ginger, etc. in order to help promote its diaphoretic effect. Nevertheless, Perilla leaf is characterised by its effects in regulating *qi*, relieving the central torso, and stopping vomiting (this effect will be promoted by combination with "zhi-qiao" i.e. Fructus Aurantii). Cases manifesting frequent nauseating and vomiting or diarrhoea may take a decoction from boiling 5 g of Perilla leaf and 3 g of coptis. Mild cold in the senile individuals or young children in whom the use of ephedra and cinnamon may cause excessive diaphoresis may be treated with Perilla leaf instead. A Chinese herbal formula named Cyperus and Perilla Formula (to be elaborated later) is one of the formulas containing Perilla as one of their component herbs.

Perilla is also good for the vomiting, pectoral distress, nausea, and lower abdominal pain experienced during pregnancy. Because Perilla stalk possesses stomachic effect, it can present antiemetic effect against pregnancy nausea and vomit, can soothe *qi* and pacify the fetus. For such purposes 4–9 g of Perilla stalk together with citrus rind and cardamon is often used to augment the stomachic effect.

Also, poisoning from eating fish and shell fish, with the symptoms of vomiting, diarrhoea and abdomen ache can be treated by taking 30–60 g of Perilla leaf decocted alone or together with fresh ginger.

Besides, Perilla can also be applied externally for treating scrotal eczema wherefore 30 g of Perilla leaf is boiled in water, and after getting cool the decoction is used to wash the lesion and then wipe the lesion with peanut oil (Chungshan Medical College, 1979).

Clinical application of Perilla seed mainly makes use of its *qi* descending and antiasthmatic effects. The herb is thus usually indicated for dyspnea, pectoral distress, wheezy stridor which in severe condition may necessitate a sitting up respiration or is accompanied by respiratory tract disturbance symptoms such as cough as seen in chronic bronchitis, pulmonary emphysema, etc. Excessive phlegm may hinder the respiratory tract passage and cause cough, dyspnea, and so forth. In traditional Chinese medicine, it is believed that eliminating phlegm may eliminate cough, dyspnea and pectoral distress which are symptoms due to pulmonary *qi* adversity, and this effect is known as descending *qi* (downing *qi*). Clinical experiences show that despite the above mentioned effects of Perilla seed, it is still necessary to incorporate Perilla seed with antitussive and expectorant herbs such as peucedanum, pinellia, etc. and *qi* regulating herbs such as magnolia bark, citrus rind, etc. in order to bring about the expected result. A Chinese herbal formula named Perilla Fruit Combination (to be elaborated later) is one of the formulas containing Perilla seed as one of the component herbs. In using the Perilla seed-containing formulas, it should be noted that because Perilla seed possesses an intestine lubricating (bowel moving) effect, it is contraindicated in patients with muddy stool or diarrhoea symptom. It can only be used in conditions showing cough with constipation (Chungshan Medical College 1979).

COMMONLY USED CHINESE HERB FORMULAS THAT CONTAIN PERILLA

As mentioned above, Perilla is often used together with other Chinese herbs in many herb formulas, especially in the *qi* formulas used for treating neurotic disorders, and respiratory diseases. In addition, it is also commonly used as a diaphoretic for common cold. Some commonly used Chinese herb formulas that contain Perilla leaf are shown

Table 1 Commonly used traditional Chinese herb formulas that contain Perilla leaf

Formula	Source	Number of Herbs	Content (%) of Perilla Leaf
Pinellia and Magnolia Combination	Jin-gui-yao-lue	5	10.0
Ephedra and Magnolia Combination	Wai-tai-mi-yao	7	7.5
Cyperus and Perilla Formula	Tai-ping-hui-min-he-ji-ju-fang	5	15.0
Ginseng and Perilla Combination	Tai-ping-hui-min-he-ji-ju-fang	13	4.4
Dang-guei Sixteen Herbs Combination	Wan-bing-hui-chun	16	5.3
Aquilaria and Perilla Formula	Tai-ping-hui-min-he-ji-ju-fang	11	9.8
Citrus and Perilla Combination	Tai-ping-hui-min-he-ji-ju-fang	15	7.7
Lindera and Cyperus Formula	Yi-xue-ru-men	6	10.7
Cyperus, Perilla and Citrus Formula	Wan-bing-hui-chun	10	9.8
Ephedra and Cimicifuga Combination	Tai-ping-hui-min-he-ji-ju-fang	12	10.2
Apricot Seed and Perilla Formula	Wen-bing-tiao-bian	11	7.6
Areca Seed and Chaenomeles Formula	Shi-fang-ge-kuo	7	5.9
Hoelen, Atractylodes and Areca Combination	Zheng-zhi-zhun-sheng	13	3.6
Agastache Formula	Tai-ping-hui-min-he-ji-ju-fang	13	4.5

Table 2 Chinese herb formulas that contain Perilla leaf as recorded in the pharmacopoeia of PRC

Formula	Number of Herbs	Content (%) of Perilla Leaf
Xiao'er Zhibao Pills	25	3.6
Xiangsu Zhengwei Pills	15	22.5
Wushicha Granules	19	2.9
Jiusheng Powder	9	13.0
Ganmao Qingre Granules	11	5.1
Huoxiang Zhengqi Pills	11	5.3
Liuhe Dingzhong Pills	17	1.5

Table 3 Commonly used traditional Chinese herb formulas that contain Perilla seed

Formula	Source	Number of Herbs	Content (%) of Perilla Seed
Perilla Fruit Combination	Tai-ping-hui-min-he-ji-ju-fang	10	12.5
Ephedra and Morus Formula	Tai-ping-hui-min-he-ji-ju-fang	7	10.0
Ephedra and Gingko Combination	Yi-fang-ji-jie	9	8.8
Atractylodes and Cardamom Combination	Wan-bing-hui-chun	12	7.8

Table 4 Chinese herb formulas that contain Perilla seed as recorded in the pharmacopoeia of PRC

Formula	Number of Herbs	Content (%) of Perilla Seed
Juhong Pills	15	5.6
Shensu Pills	11	10.7

in Tables 1 and 2. And some commonly used Chinese herb formulas containing Perilla seed or fruit are shown in Tables 3 and 4.

The so-called *qi* formulas are those used for resolving depression, descending *qi* adversity, and replenishing *qi* or for symptoms such as epigastric distension, swelling pain, flatulence, vomiting, nausea, vomiting up acid fluid, *qi* adversity, and asthma. In all, formulas capable of regulating the *qi* functions and treating the various *qi* problems are known as *qi* formulas. In the following are given a few representative *qi* formulas that contain Perilla leaf or Perilla seed as a component herb.

Pinellia and Magnolia Combination

Formula composition: pinellia 6.0 g, magnolia bark 3.0 g, hoelen 5.0 g, fresh ginger 4.0 g, Perilla leaf 2.0 g.

This formula was first recorded in a Chinese medical classic named "Jin Gui Yao Lue" (Prescriptions from the Golden Chamber) which was written in the Han dynasty of China about 1800 years ago. The formula is a representative *qi* formula which the ancients used for treating the so-called "plum kernel *qi*" (a symptom marked by a sensation in the throat where it feels as if something like a plum kernel or a piece of grilled meat is clogged there) and has been applied to the treatment of various *qi* problems (neurosis).

The formula has the function of relieving mental depression and is thus suitable for those with gastrointestinal asthenia, feeble and lax skin and muscles, mild intestinal tympanites, a sensation of gastric distension, and water stagnated in the stomach. In these patients their pulses are usually floating and weak or demersal and weak. People with such a constitution are mostly very careful, easily becoming sullen, depressed, languid, somatically weak and fatigable. In this formula, Perilla leaf possesses a mild excitant effect capable of relieving mental depression and activating the gastrointestinal functions to approach a vigorous state.

This formula is indicated for neurosis, neurasthenia, hysteria, nervousness-associated insomnia, phobia, neurotic esophageal stenosis, paroxysmal cardiac hyperfunction, esophageal spasms, bronchitis, hoarseness after a cold, asthma, pertussis, pregnancy vomit, climacteric syndrome, gastroptosis, gastric laxity, and edema (Hsu *et al.*, 1980).

Ephedra and Magnolia Combination

Formula composition: ephedra 5.0 g, licorice 2.0 g, apricot seed 4.0 g, Perilla leaf 1.5 g, magnolia bark 3.0 g, bupleurum 2.0 g, citrus rind 2.5 g.

This formula was originally recorded in the "Wai Tai Mi Yao" (An Extraminister's Secret Formulas) written by Wang Tao (675–755 A.D., Tang dynasty). It treats chronic cough which at onset manifests panting asthma causing the patient unable to sit or lie down and producing incessant wheezes in the throat followed by asphyxia. In the original text, this formula did not contain magnolia bark and licorice which were later added by Japanese herbal physicians to augment the therapeutic effects according to their clinical experiences. The formula is used in those with dyspnea as the chief complaint, and scanty phlegm associated with bronchial asthma, showing *qi* depression. Usually the formula is used for the target symptoms of weak abdominal strength, and a not very tense condition in the lower torso in patients who show only slight chest and hypochondriac distress, not much phlegm, dyspnea along with neurosis. The Perilla leaf contained in this formula possesses the effects of eliminating wind and cold evils, and, acting together with magnolia bark, descending *qi*.

This formula is indicated for bronchial asthma, pediatric asthma, and pulmonary emphysema (Hsu *et al.*, 1980).

Perilla Fruit Combination

Formula composition: Perilla seed 3.0 g, pinellia 4.0 g, citrus rind 2.5 g, magnolia bark 2.5 g, peucedanum 2.5 g, cinnamon twig 2.5 g, angelica (Radix Angelicae sinensis) 2.5 g, jujube 1.5 g, licorice 1.5 g, fresh ginger 1.5 g.

The formula is recorded in the "Tai Ping Hui Min He Ji Ju Fang" (Taiping Folks Beneficiary Dispensatory) completed in the Song dynasty of China, which is equivalent to the "Zi su zi tang" (Perilla seed decoction) recorded in the "Qian Jin Fang" (Formulas Worth One Thousand Pieces of Gold). This formula can treat chilling in the lower limbs, discontinued respirations, and dyspnea, which are often seen in the physically asthenic or senile individuals who usually manifest the symptoms of adynamia in the lower torso (below the umbilicus), oliguria, copious phlegm, short breaths with up-flushing, a stringy tense pulse that appears surging and big but forceless, and subcardiac depression. In the formula, Perilla seed, peucedanum, magnolia bark, citrus rind, pinellia and cinnamon twig all act to descend the *qi* upward adversity. Also, once the *qi* gets soothed, the phlegm is also soothed and unstagnated and hence the formula is also expectorant.

The formula is indicated for chronic bronchitis, asthmatic bronchitis, pulmonary emphysema, tinnitus, hematemesis, nosebleed, alveolar pyorrhea, oral erosion, oral cancer, edema, and beriberi (Hsu *et al.*, 1980).

Formulas capable of rendering diaphoresis, resolution of muscles and completion of eruption are known as sudorific formulas or diaphoretics which make use of the diaphoretic and muscle resolving effects to expel the pathic evil out of the surface (skin) or muscles where the evil has just invaded. Diaphoretic formulas mostly have a pungent taste, a volatile and mildly emanative property, and should be decocted gently instead of excessively, otherwise the drug quality will be lost and the efficacy reduced. Meantime, after taking a diaphoretic formula, one should avoid wind drafts or cold things, increase clothing or quilts in order to augment perspiration. The perspiration so induced should better be only to such an extent that the body is just wetted but not thoroughly wetted out. Either an incomplete perspiration or a perspiration that causes copious sweating to make the whole body soaked with dripping sweats is not the right way of rendition of perspiration, because the former condition cannot drive out the evil completely, while the latter condition can drive the evil out completely though, it is overdone and in that way it also has exhausted the primordial *qi*. A few more commonly used diaphoretic formulas containing Perilla leaf or Perilla seed are given below.

Cyperus and Perilla Formula

Formula composition: cyperus 3.5 g, Perilla leaf 1.5 g, citrus rind 3.0 g, licorice 1.0 g, fresh ginger 1.0 g.

This formula also comes from the "Tai Ping Hui Min He Ji Ju Fang". It is a representative diaphoretic *qi* formula especially good for treating *qi* stagnation and surface evil manifesting *qi* symptoms. It is suitable for those afflicted with a cold that is accompanied by both *qi* and food stagnation and for a cold in those with a weak stomach, or for symptoms induced by *qi* depression and food stagnation. The target symptoms of this formula include a demersal pulse, subcardiac distension, shoulder ache and pain, headache, vertigo, tinnitus, nausea, and *qi* stagnation. In the formula Perilla leaf acts as a diaphoretic capable of dispersing surface evils and improving blood circulation, and in particular it possesses a therapeutic effect for poisoning from eating fish.

This formula is indicated for common cold, neurasthenia, neurotic abdomen ache, climacteric syndrome, menopause, fish poisoning, urticaria, neurosis, anosmia, and stuffy nose. (Hsu *et al.*, 1980). There are many modified formulas derived from this formula.

Ginseng and Perilla Combination

Formula composition: Perilla leaf 1.0 g, platycodon 2.0 g, "zhi-qiao" (Fructus Aurantii) 1.0 g, citrus rind 2.0 g, pinellia 3.0 g, hoelen 3.0 g, pueraria 2.0 g, peucedanum 2.0 g, ginseng 1.5 g, costus root 1.0 g, licorice 1.0 g, jujube 1.5 g, fresh ginger 1.5 g.

This formula is included in the "Tai Ping Hui Min He Ji Ju Fang". It is a formula with surface-internal dual resolution effect suitable for treating common cold in any season, fever, headache, cough with water stagnancy, or internal injury by food and drink, gastric obstructive distension, vomiting, and nausea. Patients with the above conditions are usually weak constitutioned who are not suitable for treatment with ephedra-containing formulas or Pueraria Combination. Hence, the formula is often used in small children, senile individuals, asthenic people, and pregnant women who have a cold with cough. In this formula the Perilla leaf together with pueraria and peucedanum possesses a carminative effect.

The formula is indicated for common cold, bronchitis, pneumonia, *qi* impediment, pregnancy vomit, alcohol intoxication (Hsu *et al.*, 1980).

Ephedra and Morus Formula

Formula composition: ephedra 4.0 g, morus root bark 2.0 g, Perilla seed 2.0 g, apricot seed 4.0 g, hoelen 5.0 g, citrus rind 2.0 g, licorice 1.0 g.

This formula is listed in the "Tai Ping Hui Min He Ji Ju Fang" as the one formula among all ephedra-containing formulas that is most suitable for a deficiency conformation (symptom complex). It is good for treating contractions of wind and cold evils in the lungs, coughing with up rushing of *qi*, pectoral distress, spasms in the nape and back, stuffy nose with low heavy voice, dizziness and vertigo, nonfluent discharge of phlegm, a floating and quick pulse and gastrointestinal asthenia with loss of appetite. Although this formula is similar to Ginseng and Perilla Combination, yet unlike the latter which is used mostly in the senile, this formula is used mostly in small children. In this formula Perilla seed as well as hoelen, citrus rind and morus root bark is a *qi* regulating herb capable of augmenting antitussive, expectorant and antiasthmatic effects.

The formula is indicated for common cold, coughing, asthma, stuffy nose, bronchitis, pediatric asthmatic cough (Hsu *et al.*, 1980).

Both Perilla leaf and Perilla seed are important drug materials. Though the distinct uses with the two herbs have been mentioned in Chinese medical classics, there remain many problems requiring investigation in terms of their clinical application.

CONCLUSION

As a conclusion, in traditional Chinese medicine Perilla leaf is believed to be capable of rendering perspiration, resolving fever, moving *qi*, relieving the centres and antidoting poisoning from eating fish and crab, and is therefore indicated for common cold due to contraction of winds and cold, cough, asthma, pectoral and abdominal distension, tranquillisation, and stomach strengthening. Perilla stalk is believed to possess the effects of soothing *qi*, stabilising the embryo and thus can be used for treating *qi* adversity, abdomen ache, and embryonic aching and instability. Perilla seed or fruit possesses the effects of descending *qi*, arresting asthma, stopping cough, resolving phlegm, relieving the chest and alleviating depression and is thus commonly used for treating cough due to *qi* adversity, phlegmatic stridor, *qi* impediment and constipation. However, in Chinese herbal medicine these herbal articles are usually used in combination with other herbs in a prescription and are seldom used alone.

REFERENCES

Chungshan Medical College (authorship), Kobe Chuigaku Kenkyukai (translator, 1979) *Clinical Applications of Chinese Herb Drugs*, Ishiyaku Publishing Co., Ltd., Tokyo, Japan, pp. 23-24, 473 (in Japanese).

Hsu, H.Y., Chen, Y.P., Sheu, S.J., Hsu, C.S., Chen, C.C. and Chang, H.C. (1986) *Oriental Materia Medica – a concise guide*, Oriental Healing Arts Institute, Long Beach, Ca., U.S.A., pp. 59-60, 721.

Hsu, H.Y., Chen, Y.P., Sheu, S.J., Hsu, C.S., Chen, C.C. and Chang, H.C. (1985) *Oriental Materia Medica – a concise guide*, Modern Drugs of Taiwan Press, Taipei, Taiwan, pp. 37, 560-561 (in Chinese).

Hsu, H.Y. and Hsu, C.S. (1980) *Commonly Used Chinese Herb Formulas with Illustrations*, Oriental Healing Arts Institute, Los Angeles, Ca., U.S.A., pp. 41–42, 47, 126–127, 395–396, 401–402, 405-406.

Hsu, H.Y. and Hsu, C.S. (1980) *Commonly Used Chinese Herb Formulas with Illustrations*, Modern Drugs of Taiwan Press, Taipei, Taiwan, pp. 28–29, 51, 94–95, 300–302, 318–319, 328–329 (in Chinese).

Li, S.Z. (1578), reprint (1966) *Ben-cao-gang-mu*, Vol. 1, Wenkuang Book Co., Taipei, Taiwan, pp. 536–539 (in Chinese).

Namba, T. (1980) *Colored Illustrations of WAKAN-YAKU, Vol. 1*, Hoikusha Publishing Co., Ltd., Osaka, Japan, pp. 284–285 (in Japanese).

Pharmacopoeia Commission of PRC (1990) *Pharmacopoeia of the People's Republic of China*, People's Hygienic Press, Beijing, PRC, pp. 304–306 (in Chinese).

Pharmacopoeia Commission of PRC (1992) *Pharmacopoeia of the People's Republic of China*, English Edition), Guangdong Science and Technology Press, Guangzhou, PRC, pp. 279, 291, 296, 298, 299, 301, 330, 354, 355, 359, 362.

Society of Japanese Pharmacopoeia (1991) *Commentaries to the Japanese Pharmacopoeia*, 12th ed., Hirokawa Publishing Co., Ltd., Tokyo, Japan, pp. D566–D569 (in Japanese).

Takagi, K., Kimura, M., Harada, M. and Otsuka, Y. (1982) *Pharmacology of Medicinal Herbs in East Asia*, Nanzando Co., Ltd., Tokyo, Japan, pp. 216–217 (in Japanese).

5. ANTI-INFLAMMATORY AND ANTIALLERGIC ACTIVITIES OF PERILLA EXTRACTS

MASATOSHI YAMAZAKI and HIROSHI UEDA

Department of Medicinal Chemistry, Faculty of Pharmaceutical Sciences, Teikyo University, Sagamiko, Tsukuigun, Kanagawa 199-01, Japan

INTRODUCTION

Many types of cells, including macrophages, produce tumor necrosis factor (TNF). TNF was recently recognized as an important host defense factor that affects not only tumor cells but many kinds of normal cells (Vilcek and Lee, 1991). However, prolonged exposure to TNF might contribute to the wasting of the host which is associated with many chronic disease states (Beutler, 1988). Moreover, acute overproduction of TNF in bacterial infection may induce septic shock leading to acute organ failure and death. Inhibiting overproduction of TNF is therefore an advantageous step toward the suppression of acute and chronic inflammation. This work was undertaken to ascertain whether Perilla extract, which may be a new candidate for an antiinflammatory and antiallergic reagent, can inhibit TNF production and inflammatory states in mice.

MATERIALS AND METHODS

Animals and Cell Line

Males of specific pathogen free strains of C3H/He and ICR mice were used at 6–8 weeks of age. L-929, a transformed fibroblast cell line originally derived from a C3H/He strain mouse, was used for TNF assay.

Culture of Macrophages

Peritoneal exudate cells were obtained from mice 24 hrs after *i.p.* injection of 4 mg glycogen. These cells were suspended in RPMI-1640 medium supplemented with 5% heat-inactivated fetal bovine serum. The peritoneal cells were incubated in 96-well microtest plates for 1.5 hr at 37° C and the medium was removed to obtain adherent cells. More than 95% of these adherent cells were macrophages, as determined by Giemsa staining and measurement of uptake of carbon particles. These exudative macrophages were finally cultured in 0.2 ml of medium with or without test samples. The TNF concentration of macrophage supernatant was measured immediately after its collection.

Perilla Extracts

Dried leaves of green types of Perilla (*Perilla frutescens* (L.) Britton var. acuta Kudo forma virides Makino) were mixed with an equal volume of distilled water. This mixture was

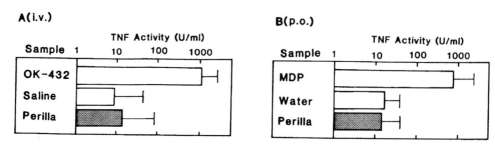

Figure 1 Priming effect of Perilla extracts on TNF production.

homogenized with a polytron and centrifuged to obtain supernatant. The supernatant was orally administered. The extracts were adjusted to pH 7.4 and to an isotonic condition with an osmometer, and then sterilized with a millipore membrane filter to use intravenously. Some of Perilla extracts were kindly provided by Amino Up Chemical Co., Ltd. (Sapporo, Japan).

Measurement of TNF Activity

Production of circulating TNF was measured by the L cell cytotoxicity method. Briefly, TNF activity of serum preparation was measured by *in vitro* 18-hour cyto-toxicity assay in the presence of actinomycin D. Recombinant human TNF was used as a standard TNF preparation. The specificity of TNF in this assay was checked by an antibody-neutralization test using anti-mouse TNF rabbit antibody. In this study, OK-432 (an antitumor reagent from a Gram-positive bacterium, *Streptococcus pyogenes*) was chosen as the TNF-trigger, because of its stable triggering effect and potent activity comparable to LPS, but with no acute toxicity to animals.

INHIBITION OF TNF PRODUCTION BY PERILLA EXTRACTS *IN VIVO*

There are two steps in the process of producing TNF: a priming step (e.g.,cytokine or immunopotentiator pretreatment) and a triggering step (e.g.,bacteria or LPS treatment). We previously reported that a primed state ready for TNF triggering can be obtained by several immunopotentiators and some selected cytokines such as interferons and interleukin-2 (Okutomi and Yamazaki, 1988). We also showed that some vegetable extracts can prime the endogenous production of TNF *in vivo* (Yamazaki *et al.*,1992). Here, we examined whether Perilla extracts are capable of causing this priming effect for TNF production.

   As shown in Figure 1, the TNF activity triggered by OK-432 (3KE) without a priming sample was below 20 U/ml. After priming with MDP (0.5mg) or OK-432 (0.3KE) with triggering by OK-432 (3KE), TNF production was significantly enhanced to about 1000 U/ml. Perilla extracts, however, could not induce the priming effect.

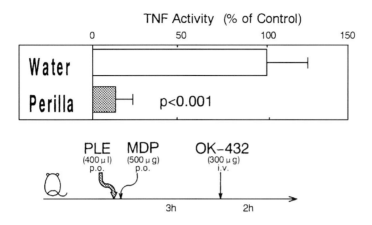

Figure 2   Inhibition of endogeneous TNF production by Perilla.

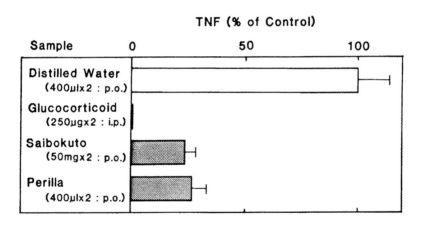

Figure 3   TNF-inhibitory activities of anti-inflammatory reagents.

Next, we examined whether Perilla extracts inhibit the endogenous production of TNF *in vivo*. Mice were administered *p.o.* 0.5 mg of MDP and 3 hrs later were given *i.v.* injection of OK-432. These mice produced 1000-3000 U/ml of TNF. Perilla extracts were administered *p.o.* simultaneously with MDP. As shown in Figure 2, TNF activity was significantly inhibited by *p.o.* administration of green Perilla extracts. Glucocorticoid is a powerful antiinflammatory reagent and saibokuto, a Chinese medicine complex, is also widely used to treat inflammatory disease in Japanese clinics. Figure 3 shows that Perilla extracts have nearly the same activity against TNF production as does saibokuto.

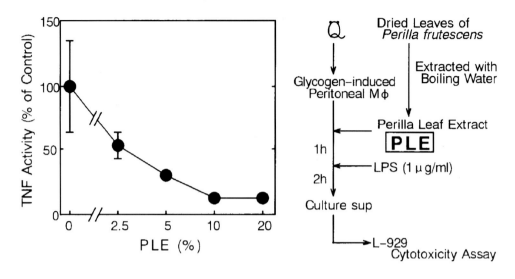

Figure 4   Inhibition of TNF production from macrophages by Perilla

INHIBITORY ACTIVITY OF PERILLA EXTRACTS FOR *IN VITRO* TNF PRODUCTION

Macrophages can release TNF into a culture after appropriate stimulation *in vitro*. We examined whether Perilla extracts can also inhibit TNF production from macrophages *in vitro*. For this, glycogen-induced macrophages were incubated with Perilla extracts for an hour, then LPS-triggered TNF release for 2 hours was measured. As shown in Figure 4, Perilla extracts inhibited TNF-release from macrophages *in vitro* as well as *in vivo*. Pretreatment of macrophages was effective even 10 min after addition of the extracts.

ANTI-INFLAMMATORY ACTIVITIES OF PERILLA EXTRACTS

Certain irritants can induce acute inflammation in mice. Phorbor ester (TPA) induces leukotriene-dependent inflammation and arachidonic acid induces prostaglandin-dependent inflammation. We selected these two typical irritants as acute inflammatory models. Ear swelling of mice was maximally observed 2–4 hrs after painting of these irritants.

Perilla extracts (0.4ml/mouse) were administered *p.o.* 3 hrs and 18 hrs before the painting of 0.25mg of arachidonic acid. As shown in Figure 5, the extracts inhibited arachidonic acid-induced ear swelling of mice. However, their administration after arachidonic acid painting was ineffective. TPA-induced acute inflammation was also inhibited by *p.o.* administration of Perilla extracts (Figure 6), but was not inhibited by their post-treatment. Perilla extracts suppressed two typical acute inflammations.

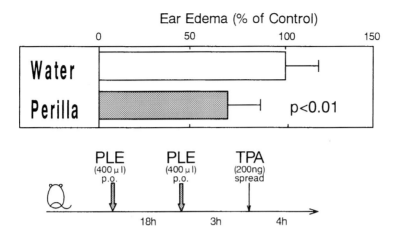

Figure 5   Inhibition of TPA-induced ear swelling by Perilla

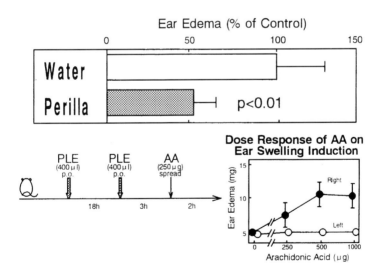

Figure 6   Inhibition of arachidonic acid-induced ear swelling by Perilla

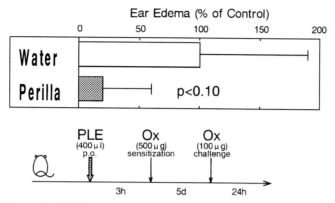

Figure 7   Inhibition of oxazolon-induced ear swelling by Perilla

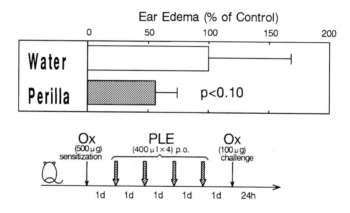

Figure 8   Inhibition of oxazolon-induced ear swelling by Perilla

ANTI-ALLERGIC ACTIVITY OF PERILLA EXTRACTS

Oxazolon induces the type IV allergy. For sensitization, oxazolon was painted onto the abdominal skin of mice and 5 days later challenge-painting was done. Ear swelling was checked 24 hrs after oxazolon challenge. Perilla extracts inhibited oxazolon-induced ear edema when administered before sensitization (Figure 7). Although this delayed type allergy was also suppressed by *p.o.* administration of Perilla extracts prior to challenge (Figure 8), the suppression was not observed after oxazolon challenge.

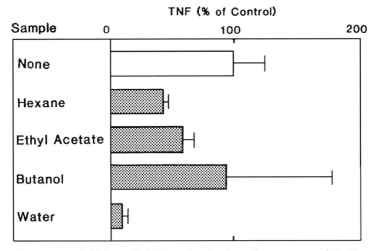

Figure 9   Inhibition of TNF production by solvent-extracted Perilla

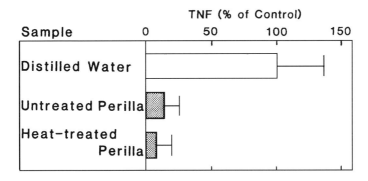

Figure 10   Heat stability of inhibitory effect of Perilla on TNF production

CHARACTERIZATION OF PERILLA EXTRACTS

To identify the active ingredients in Perilla extracts, we first focused on flavor substances of *Perilla frutescens*, which contains various kinds of flavor chemicals, the main perfume substance being perillaldehyde. Perillaldehyde did not inhibit the endogenous production of TNF.

Next, we extracted the active substances in *Perilla frutescens* using several solvents. As shown in Figure 9, the inhibitory activity for TNF production was largely recovered in the water-soluble fraction but not in the fraction of butanol. Both hexane and ethyl acetate fractions showed relatively weak activity. These results suggest that most of the active ingredients in Perilla extracts are hydrophilic rather than hydrophobic substances.

The active substances inhibiting TNF production was heat stable, showing no loss of activity after heating at 100°C for 10 min (Figure 10). The inhibitory activity was restored in the fraction of molecular size below 10,000 daltons. From these results the active factors in *Perilla frutescens* are recognized to be stable and small sized molecular substances.

CONCLUSION

TNF is an important host defense factor that affects many kind of normal cells. (Vilcek and Lee, 1991). Its overproduction, however, is associated with acute and chronic inflammation and allergy (Beutler, 1988). Anti-TNF antibody was reported to suppress both types III and IV allergies (Piguet *et al.*,1991; Zhang *et al.*,1992). TNF is produced from many types of cells including mast cells which are key cells in inflammation and allergy (Walsh *et al.*,1991). TNF can also stimulate IgE production. The suppression of TNF production is thus a clue to the causes of inflammation and allergy.

Here, we reported that Perilla extracts inhibited TNF production both *in vivo* and *in vitro*. Oral administration of these extracts suppressed TPA- and arachidonic acid-induced inflammation and oxazolon-induced allergy. The active substances in *Perilla frutescens* may be hydrophilic, stable and small molecular compounds. Perilla extracts may thus be a new candidate for an anti-inflammatory and antiallergic reagent.

REFERENCES

Beutler, B. (1988) Tumor necrosis,cachexia,shock and inflammation: a common mediator. *Ann. Rev.Biochem.*, **57**, 505–518.

Okutomi, T. and Yamazaki, M. (1988) Augmentation of release of cytotoxin from murine bone marrow macrophages by IFN-r.*Cancer Res.*, **48**, 1808–1811.

Piguet, P.F., Grau, G.E., Hauser, C. and Vassali, P. (1991) Tumor necrosis factor is a critical mediator in hapten-induced irritant and contact hypersensitivity reactions. *J. Exp. Med.*, **173**, 673–679.

Vilcek, J. and Lee, T.H., (1991) Tumor necrosis factor. *J. Biol. Chem.*, **266**, 7313–7316.

Walsh, L.J., Trinchieri, G., Waldorf, H.A., Whitaker, D., and Murphy, G.F. (1991) Human dermal mast cells contain and release tumor necrosis factor, which induces endothelial leukocyte adhesion molecule 1. *Proc. Natl. Acad. Sci. USA.*, **88**, 4220–4224.

Yamazaki, M., Ueda, H., Fukuda, K., Okamoto, M. and Yui, S. (1992) Priming effects of vegetable juice on endogenous production of tumor necrosis factor. *Biosci. Biotech. Biochem.*, **56**, 149.

Zhang, Y., Ramos, B.F. and Jakschik, B.A. (1992) Neutrophil recruitment by tumor necrosis factor from mast cells in immune complex peritonitis. *Science*, **258**, 1957–1959.

6. PERILLA AND THE TREATMENT OF ALLERGY— A REVIEW

HE-CI YU, AIMO NISKANEN and JUKKA PAANANEN

Hankintatukku Natural Products Co.
Temppelikatu 3-5 A5, Helsinki, Finland 00100

Perilla (*Perilla frutescens* Britt.), a traditional Chinese herb, has recently received special attention because of its beneficial effects in the treatment of some kinds of allergic reactions without the side effects associated with some other used antiallergy medicines. In this chapter, the authors present a review of the problem of allergy and the current favorable evidence for the use of Perilla products towards its resolution.

THE ALLERGY PROBLEM

Allergy is an abnormal immune reaction of the body to allergens such as pollen, dust, certain foods, drugs, animal fur, animal pets, animal excretions, feathers, microorganisms, cosmetics, textiles, dyes, smoke, chemical pollutants and insect stings. Certain conditions such as cold, heat, or light may also cause allergic symptoms in some susceptible people. Some allergens are just specific to some individuals but not to others. Allergens may act via inhalation, ingestion, injection or by contact with the skin. The resulting allergy may cause the victim to have a medical problem such as hay fever (allergic rhinitis), or atopic dermatitis (eczema), or allergic asthma, with symptoms ranging from sneezing, rhinorrhea, nasal itch, obstruction to nasal air-flow, loss of sense of smell, watery and itching eyes as in allergic rhinitis; and skin itching, skin redness and skin lesion as in atopic dermatitis. Allergic patients suffer not only irritating symptoms but also an impairment of the quality of life.

In 1989 allergy diseases affected 10% of the world's population (Vercelli and Geha, 1989). Today, allergy is even more common. It is now the most wide spread immunological disorder in humans, and the most prevalent and rapidly increasing chronic health problem, particularly in the industrialized countries where allergy affects one in four individuals. Furthermore allergy also causes a socioeconomic problem resulting in huge economic losses. In the U.S. alone it costs billions of dollars per year (HayGlass, 1995).

In the U.S. an estimated 20% or more of its population is allergic to something (Lichtenstein, 1993). Allergic rhinitis alone affects 10–15% of the U.S. population (Broide, 1995). The incidence of allergic asthma in the U.S. rose by 60% between 1979 and 1989 (Bousquet *et al.*, 1994). In England an investigation showed that the reported prevalence of asthma had risen from 4.1% in 1964 to 10.2% in 1989, hay fever from 3.2% rose to 11.9%, and eczema from 5.3% rose to 12% in the same period (Rusznak *et al.*, 1994). In Central Europe, pollinosis alone affects 10 percent of its population (Puls and Bock, 1993). In Sweden, more than one in three adults and 40% of children suffer from allergy, and the number of allergy patients has doubled over the last ten years and continues to increase (Ullenius *et al.*, 1995). In Finland, 20–30% of young adults have the allergic symptoms (rhinorrhea), and about 20% of aged 0 to 6 year children have the skin eruption symptom (Nuutinen,1995; Syvänen, 1995). In Australia, asthma affects as many as 20% of children and 10% of adults (Sutherland, 1994). In Japan, an investigative data published by the Ministry of Welfare of Japan showed 34% of its population suffered from some kinds of allergies, and mostly among children from the age of 0 to 4 (Oyanagi, Chapter 7, this book).One-third of infants born in Japan are diagnosed as atopic (Okuyama, 1992).

One main cause which contributed to the increase in allergy incidence is the change of living environment and the extension of air pollution which is becoming more and more serious all over the world. In Japan, allergic rhinoconjunctivitis was found to be more prevalent in individuals living near motorways than in cedar forests. A recent study in Finland showed that admissions to hospital with severe asthma correlated with atmospheric levels of nitrogen dioxide (Rusznak *et al.*, 1994). In China, the prevalence of asthma is greater in the more modernized area than in the less developed area (Zhong, 1994).

On the other hand, as a result of the affects of air pollution, the climatic change of global warming is likely to have significant effects on the distribution and abundance of allergenic plants. This impact on natural ecosystems is likely to be greatest in southern and Mediterranean Europe and in northern Scandinavia. These changes will alter the severity of pollen seasons and will have wide ranging implications for the incidence of pollinosis (Emberlin, 1994).

However, the prevalence of allergy in China is generally much lower than in western countries. Hay fever, the leading allergic disease in the West, is not common in China and some other Asian countries (Zhong, 1994; Leung, 1993; Rubenstein and Rubenstein, 1984). It is interesting to investigate whether this difference is due to racial predisposition or environmental factors; or may partly be attributed to the popularity of the application of traditional Chinese medicines in China.

THE TREATMENT OF ALLERGY

The treatment of allergic diseases has benefited from the gradual understanding of allergy mechanisms. Allergic reaction is due to a change in the immunoreactivity of an individual. From current knowledge, its mechanism (Lichtenstein, 1993; Kay, 1993; International Rhinitis Management Working Group, 1994a; Cooper, 1994), is partly illustrated in Figure 1. When the allergen is presented to T-helper cells, B-lymphocytes will overproduce allergen-specific IgE antibodies. In the case of allergen contact, IgE bound to mast cells leads to mast cell activation and degranulation. As a result, mast cells release abnormal amounts of mediators such as histamine, PAF (platelet-activating factor), leukotrienes, prostaglandin D. These mediators will dilate blood vessels, increase permeability of small blood vessels, stimulate nerve endings, stimulate secretion of mucus in airways, or constrict bronchial airways, so as to induce local inflammation and cause various immediate symptoms or chronic symptoms.

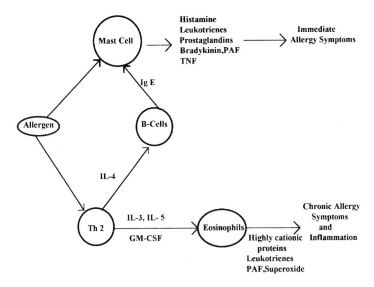

GM-CSF : Granulocyte-Macrophage Colony Stimulating Factor
Ig E : Immunoglobulin E
IL- : Interleukin
PAF : Platelet-Activating Factor
Th 2 : T helper lymphocyte 2
TNF : Tumor Necrosis Factor

Figure 1   Hypothesis on Partial Mechanisms of Allergy

In addition, cytokines such as TNF (Tumor Necrosis Factor) were recently found to be linked with allergic reaction. Indeed, plasma TNF levels are elevated in the serum of patients with atopic dermatitis, and the levels are tightly correlated with plasma histamine (Cooper, 1994; Sumimoto *et al.*, 1992). Furthermore, TNF will reversely stimulate the immune cells to produce more mediators (Yamazaki, 1993). It is now accepted that, allergy pathogenesis is involved in multifactors including hereditary, environmental, and immunological factors.

The treatment of allergic diseases is mainly based on allergen avoidance, public health education, the use of chemotherapy (therapy using any synthetic pharmaceutical compound), traditional herbal therapy, immunotherapy or an herbal immune regulator. In this chapter, Perilla is advanced as a herbal immune regulator.

Allergen Avoidance

If the allergens can be identified, allergy may be treated simply by avoiding the offending agents. Allergen avoidance has always been seen as the first choice to be tried and the most effective treatment, but preventing exposure to some common environmental substances such as house dust mite, and pets allergens, is seldom possible. As mentioned above the main cause of the allergic reaction is air pollution, therefore it is virtually impossible to prevent outdoor allergens exposure completely. Furthermore, even when an allergen, such as a pet is removed from a patient's environment, the benefit may take several weeks or months to be perceived (Bousquet *et al.*, 1994).

Public Health Education

Public health educational campaigns are helpful in promoting self-medication and rational and economic treatment of allergy. For example, in Sweden the year 1995 was proclaimed as the Allergy year by five related national organizations for improving quality of life of allergy sufferers (Ullenius *et al*, 1995) and in Finland, a similar campaign was undertaken by Finnish community pharmacists (Aaltonen and Kostiainen, 1995).

Chemotherapy

As a main measure, medicine therapy plays an important role in allergy treatment. But with today's medicines such as antihistamine, corticosteroids, sodium cromoglycate and so on , it is still symptomatic (International Rhinitis Management Working Group, 1994b; Cooper, 1994). The symptoms of allergic diseases are caused by various factors and are different in individuals, and therefore therapy must be varied accordingly. Present medicines can cause side effects such as osteoporosis, diabetes, weight gain, ulcer and resistance to the corticosteroids. Although new application techniques have dramatically decreased the unfavorable side effects, some newer effective approaches to allergy treatment without side effects are still eagerly awaited by sufferers and physicians.

Immunotherapy

Immunotherapy is a specific form of controlled allergen admission that changes immunoactivity into allergen tolerance. It has been used for more than 80 years but still represents a controversial treatment of allergic diseases (Malling, 1994). It is appreciated that the efficacy of allergen immunotherapy is currently very low. Severe symptomatic reactions occasionally may occur, especially in asthmatic patients. In certain countries (the UK and the Scandinavian countries) the use of immunotherapy has been greatly curtailed due to adverse reactions (International Rhinitis Management Working Group, 1994b; HayGlass, 1995).

Traditional Chinese Medicine (TCM)

With thousands of years of experience in treating diseases with natural materials, TCM still plays an important role in the health-care system of modern China and is officially recognized not only in China, but also in Japan and in some other eastern and south-eastern Asian countries which have the same cultural tradition (Zhu and Woerdenbag, 1995). People there still prefer to use traditional herbs for the treatment of allergy. There are several TCM prescriptions effective for allergic symptoms (Chen Yuh-Pan,

Table 1 Some traditional Chinese herb prescriptions for allergy

Condition	Prescriptions
Allergic Rhinitis and Pollinosis	Minor Blue Dragon Combination
	Ophiopogon Combination
	Pueraria Combination
	Magnolia flower and Gypsum Combination
	Ephedra and Apricot Seed Combination
	Atractylodes Combination
	Ephedra Combination
	Minor Bupleurum Combination with Pinellia and Magnolia Combination (with **Perilla**)
	Ephedra, Aconiti and Asarum Combination
Bronchial Asthma	Minor Blue Dragon Combination
	Minor Bupleurum Combination with Pinellia and Magnolia Combination (with **Perilla**)
	Ephedra and Apricot Seed Combination
	Minor Bupleurum Combination with
	Ephedra and Apricot Seed Combination
	Ephedra and Magnolia Combination (with **Perilla**)
	Hoelen and Schizandra Combination
Urticaria	Minor Blue Dragon Combination
	Pueraria Combination
	Bupleurum and Schizonepeta Combination
Atopic Dermatitis	Zemaphyte
Crab and fish poisoning	**Perilla** and Ginger Combination

Personal communication). Some of them are listed in Table 1. Among them, the Minor-Bupleurum-Combination with Pinellia-Magnolia Combination (Saibokuto), consisting of ten herbs including Perilla and the Ephedra-Magnolia-Combination containing Perilla, have been traditionally used for the treatment of bronchial asthma. Studies showed that Saibokuto, and Minor-Blue-Dragon-Combination (Syoseiryuto) clinically exhibit inhibitory effects on type I allergic reaction (Umesato *et al.*, 1984) and Saibokuto was found to inhibit histamine release and mast cell degranulation (Toda *et al.*, 1988).

Furthermore, there has been increasing interest in western countries for TCM. British scientists achieved impressive results with a TCM prescription (Zemaphyte) containing 10 Chinese herbs for the treatment of severe atopic eczema (Latchman *et al.*, 1995; Sheehan and Atherton, 1994; Cooper, 1994; Sheehan *et al.*, 1992; Harper *et al.*, 1990). Dr. Allan indicated in the FIP-CPA'93 (International Pharmaceutical Federation–Chinese Pharmaceutical Association) that most western medicines are single substances intended for a single well-defined disease. Complex diseases like eczema were not well served by western medicine but are often successfully treated with TCM (Hardman, 1993). Cooper indicated that the mechanism of action of Chinese herbal mixtures and their toxicities require further investigation, but may reveal hitherto unconsidered avenues (Cooper, 1994). Therefore, traditional herbs are a potential valuable source for obtaining effective medicines for the treatment of allergy.

Herbal Immune Regulators

In recent years, many researchers have been interested in elucidating the function of TCM as biological response modifiers or immune regulators. It has also been confirmed that some common vegetables or herbs contain nonnutritive components that may provide protection against chronic diseases including allergy and even some forms of cancer. (Haranaka *et al.*, 1985; Chen and Chen, 1989; Kawakita *et al.*, 1990; Yamazaki, 1992; Imaoka *et al.*, 1993).

A research team led by Prof. Yamazaki of Teikyo University, Japan, reported about the screening of vegetables with immune regulating activity. Experiments *in vivo* and *in vitro* found that among 18 kinds of vegetables, Perilla and ginger were the most active in reducing TNF production and its activity, which is linked with the allergy and inflammation as mentioned above (Yamazaki, 1992). Based on these findings, Kosuna and Yamazaki have developed the new application of Perilla in the treatment of allergy (Kosuna, 1993). On the other hand, it has also been found that Perilla seed oil rich in n-3 fatty acid (α-linolenic acid) is also said to have some benefit in the treatment of allergy and this is dealt with later in this chapter (Ito *et al.* 1992; Watanabe *et al.*, 1994).

Interestingly, these findings agreed with earlier reports on the application of Perilla and ginger mixture for the clinical treatment of asthma and chronic bronchitis (Jiangsu New Medical College, 1977). Further reports trace back to the traditional use of Perilla leaf and seed for hundreds of years in the treatment of asthma (Chen, Chapter 4) and some symptoms associated with what is now known as allergy. Also, the traditional method of cooking crab or shellfish with Perilla leaves, in order to prevent so called "poisoning" existing in crab etc., might be re-evaluated as an effective way of preventing food allergy (IgE-mediated allergy) (Ortolani and Vighi 1995; Burks, 1995).

APPLICATION OF PERILLA LEAF EXTRACT FOR ALLERGY

The manufacturing process of Perilla extract and the safety of the products were reported by Kosuna in Chapter 8 of this book. In addition to the factory made Perilla products, several other methods for Perilla preparation are available in the folklore of China and Japan (Yamazaki, 1994; Kozo, 1994; Xu, 1983). The application of home made Perilla extract is also used for the treatment of allergy. However, the removal from the extract of agricultural chemicals and perillaldehyde, which might be allergens to some individuals is important (Kosuna, Chapter 8).

Administration of Perilla Leaf Extract

Dr. Oyanagi *et al.* reported their experiences in treating allergy patients with Perilla products (Oyanagi, 1993; Yamagata, 1992; Mitsuki, 1992; Kabaya, 1994). According to the different symptoms and the condition of the patients administration may be singly or combined (Oyanagi, 1993; Yamagata, 1992; Mitsuki, 1992).

Oral administration

For the concentrated products of Perilla extract, the dose was 0.3–2 ml/50–100 ml water or other drink, 2–3 times daily dependent on the age. With home made or diluted extract, the dosage varied with the concentration and methods of preparation.

Nasal application

To relieve the symptoms of an itching or running nose, the Perilla extract was applied inside the nostrils using a cotton bud.

Topical application of Perilla extract

Application to the skin was helpful in relieving the itching and redness.

Topical use of Perilla cream and soap

For some atopic dermatitis, Perilla cream and soap were recommended (Oyanagi, Chapter 7).

Following treatment with Perilla, some patients experienced effects just after use, others after some days to one week, or in some cases up to three months. A therapy period usually occupied three months. In some cases, at the beginning of the treatment, the symptoms seemed to be more serious than before. This reflects an action of Perilla. Usually after one week, the symptoms are less and thereafter the patient continues to improve. Only a small number of patients (about 2–3%) will experience worse symptom even after ten days. For these, the Perilla extract is discontinued (Oyanagi, 1993; Yamagata, 1992; Mitsuki, 1992).

Evaluation of Usefulness of Perilla Leaf Extracts

Perilla leaf extract has been available as a "health product" rather than as a medicine. There are no published reports of controlled clinical trials. Even so, there are many reports of open (uncontrolled) studies from physicians and from patients-completed questionnaires, to support the beneficial use of Perilla leaf extract in the treatment of allergy. Rigorous double-blind placebo-controlled trials are doubtlessly needed before Perilla leaf extract can be accepted as an antiallergy medicine in the West.

Out-patients evaluated by physicians

Dr. Oyanagi reported his results of open studies in the treatment of more than one hundred allergy cases of children with atopic dermatitis. After three months of therapy

Table 2   Conditions of 78 cases who were effective after using Perilla extracts for allergy treatment

Conditions		Cases	%
Gender	Female	52	62.8
	Male	26	37.2
Age (years)	0–10	24	30.8
	11–20	13	16.6
	21–30	12	15.4
	31–40	10	12.8
	41–50	11	14.1
	51–60	6	7.7
	61–70	1	1.3
	71–80	1	1.3
Allergic	0–3	14	17.9
history (years)	4–6	13	16.7
	7–9	6	7.7
	10–15	6	7.7
	16–30	7	9
	Unknown	32	41
Allergic	Pollinois	20	25.6
diseases	Rhinitis	21	26.9
	Atopic	36	46.2
	Asthma	1	1.3

using a Perilla extract cream formulation, 80% of the patients showed varying degrees of improvement in the degree of itching, skin lesion, and eruption (Oyanagi, 1993, Yamazaki, 1994). No side effects were observed in any of the cases.

Dr. Takiguchi (1993) reported from his open studies that 20 allergic patients, using Perilla cream topically and Perilla extract orally, after two months, showed an improvement in 90% of the patients. Among these, 30% of the patients got significant benefit, 25% of the patients got some benefit and the rest got little benefit. All these patients ceased other medicine while using the Perilla products. Dr. Takiguchi concluded that Perilla is effective for allergy treatment without side effects.

Self evaluation by patients

In the Japanese magazines "My Health" (No. 9, 156–168, 1992; No. 8, 142–145, 1993; No. 3, 192–206, 1993; No. 2, 145–149, 1994), "Anshin" (No. 7, 171–191,1992; No. 7 139–155, 1993; No. 7, 258–267, 1994) several special issues about the evaluation of the use of Perilla extract for allergy were published. Many allergic patients reported their satisfying results after using Perilla leaf extracts or its products such as Perilla cream, Perilla soap and Perilla drink. From these the following data is derived.

Seventy eight cases reported effective benefit after using Perilla extract. Two thirds of them were female and one third of them were male. They included various age groups, but nearly 50% were under the age of twenty. The allergy history of these cases, varied from months to thirty years. A strong familial tendency could be seen. Among these 78 cases, 36 cases were included in 14 families and heredity was implicated. Among the cases, about half suffered from pollinosis or allergic rhinitis, and half had atopic dermatitis. Only one case was an asthma sufferer (Table 2). Those were allergic to pollen had allergy in Spring and Summer, and some also in Autumn. With Perilla treatment all these patients got some relief from their allergy symptoms and their quality of life was significantly improved.

Few cases suffering from asthma benefited from using Perilla extract. This might be attributed to the fact that plasma TNF-α concentration in bronchial asthma cases was not increased or only slightly increased (Sumimoto et al.,1992). Whereas the action of Perilla extract was mainly due to its inhibition of TNF overproduction (See Figure 1 and Chapter 5, this book).

Since 1990 Perilla extract has been widely recommended for the treatment of allergy and some details are given from case studies:

A lady aged 30 years, had allergic reaction since infancy. Every spring when the trees were budding or in late summer when some plants were in bloom, she would have allergic rhinitis, recurrent sneezing, watery nose and eye, itching nose and eye. At that time her medication was ineffective, and she was also concerned about the side effects of the medication. Her child, aged 6, also suffered from allergic rhinitis and atopic dermatitis. After using Perilla extract only a few times, her allergic symptoms were reduced significantly. Her child was also treated topically with Perilla extract for his eczema and obtained as good results as his mother. Since Perilla extract has no side effects, they took Perilla extract as a drink (5-6 drops in a cup of water) twice daily, in order to prevent their symptoms (Mizumoto, 1992).

Another lady, aged 42 years, and a beautician, could not work very well due to pollinosis. Her history of allergy was short but the symptoms were serious with recurrent sneezing, nasal itching, face and eye itching, and running nose requiring a full box of tissues every day. After using Perilla extract for one week the running nose stopped and her quality of life was markedly improved (Inaba, 1992).

A man, aged 62 years, suffered from allergic rhinitis, associated with change in the weather, such as rain or strong wind. After using Perilla extract for three months, his allergic symptoms were much reduced (Tajiri, 1993).

Self-completed questionnaires

In Finland, Perilla extract was launched on the market in 1993. Investigation of the usefulness of Perilla products for allergy was carried out by self-completed questionnaires. These included age, gender, allergy history, symptoms, specific allergen, time of allergy, medicines used before, dosage, and evaluation of the effectiveness. The latter was self-evaluated according to the reduction of symptoms and the quality of life. The investigation is continuing in some other European countries. In one group in Helsinki city forty cases had accepted questionnaires, among them female 50 %, male 50 %.

Investigation showed all the cases were allergic to pollens, about half of the cases were also allergic to some foods or animal pets. All the cases had allergy in spring time and about 40% of the cases, had the trouble all year round. Of the allergy cases investigated, 30% reported Perilla products were highly effective, 35% reported effective, 20% less effective and 15% ineffective.

One example: A young man was allergic to dogs and cats experiencing eye itching and running nose. He took Perilla capsules, 2 or 3 capsules daily, for two weeks before his friend arrived with his dog to spend holiday with him and he had no allergic reaction during their stay (Yu, 1996).

USE OF PERILLA OIL FOR ALLERGY

Bjorneboe *et al.* (1989) reported that dietary modifications involving supplementation with fish oil (rich in n-3 fatty acids) slightly favored the experimental group over the control group in the case of atopic dermatitis. Dietary Perilla oil, rich in n-3 fatty acids (which contain 54–64% linolenic acid) was also found to have some benefit in atopic dermatitis (Ito *et al.*, 1992).

Recent studies have indicated that the dietary α-linolenate/linoleate balance affected on the development of some chronic diseases including allergy (Okuyama, 1992). Okuyama proposed that the excessive intake of linoleic acid and the changes in the essential fatty acid balance of diets have made our bodies hyperreactive to various allergens. The increase in the intake of Perilla oil will inhibit the production of leukotrienes which is another important mediator causing allergy (Figure 1). Therefore taking Perilla oil might be beneficial for the prevention and treatment of allergy.

Some recent patents have been published about the efficacy of Perilla oil for the treatment of allergy (Jpn. Patent Kokai Tokkyo Koho JP 290822, 1992; JP 290812,1990). In one study (Table 3) twenty four patients with atopic dermatitis patients (male 18, female 6, aged 5–32 years) were divided into four groups each taking different kinds of oil, 2–3 grams. The different oil groups had different ratios of n-6/n-3 fatty acids. After three months those taking Perilla oil had benefited significantly (Jpn. Patent Kokai Tokkyo Koho JP 290822, 1992). Ito *et al.* (1992) also reported a pilot study on 6 atopic dermatitis outpatients treated with Perilla oil. These patients were given α-linolenic acid-enriched diet reducing the n-6 fatty acid (linoleic acid and arachidonic acid) intake and increasing the n-3 fatty acid (Perilla oil). After 40 days treatment the dermatitis improved in 3 patients. Results suggested that daily meals containing α-linolenic acid-enriched diet may have some benefit in the treatment of allergic diseases.

THE POSSIBLE MECHANISMS OF PERILLA IN THE TREATMENT OF ALLERGY

Although the precise mechanisms of Perilla treatment for allergy are not yet well elucidated, recent researches on the various phytochemicals and their pharmacological properties have also revealed some mechanisms of Perilla action in allergy. Kosuna (1995) recently published a review on anti-inflammatory active compounds in Perilla.

Table 3 Efficacy of Perilla oil in allergy

Patient Group	Intake Oil	n6/n3	Improvement % after one month	two months	three months
1	Perilla oil 100%	0.22	33	50	83
2	Sesame 100%	105.76	17	33	33
3	Safflower oil 100%	382.5	0	0	17
4	Perilla oil 80% Sesame oil 20%	0.4	50	83	83

Several active components contained in Perilla have been found to be linked with antiallergy and anti-inflammatory actions. These include elemicine, α-pinene, caryophyllene, myristicin, β-sitosterol, apigenin, phenylpropanoids and also some flavonoids which act as anti-inflammatory agents (Martin *et al.*, 1993). From current knowledge, the mechanisms of allergy treatment by Perilla may involve the following aspects which are linked to the regulation of the condition by the immune system.

Perilla Leaf Extract

1. TNF inhibition

Relevant to this section is the Perilla leaf extract which contains active components of molecular weight less than 10000. As mentioned above, Yamazaki reported that Perilla extract was shown to be active in inhibiting TNF production (Chapter 5). Kosuna proposed that more than ten active components contained in the Perilla leaf extract were active in inhibiting TNF production which plays an important role in controlling allergic reaction (Figure 1) (Kosuna, Chapter 8, this book). Sumimoto *et al.* (1992) reported that plasma TNF-α concentration was increased in atopic dermatitis and the magnitude of the increase was correlated with the severity of the dermatitis. Also a significant correlation was found between plasma TNF-α and plasma histamine concentrations in atopic dermatitis. Therefore, it seemed to be reasonable that Perilla was useful in treatment of atopic dermatitis and some other allergic reactions. However, TNF-α concentration was not or only slightly increased in plasma from bronchial asthma patients. This may explain why Perilla extract is less effective in asthma patients.

2. IgE inhibition

For the Perilla leaf extract which contains active components with a molecular weight more than 10000, Imaoka *et al.* (1993) indicated that immunosuppressive effects of this kind of Perilla leaf extracts are preferentially on IgE production and that may be useful for the suppression of IgE antibodies in certain allergic disorders (Figure 1).

3. Antioxidative activity

As reported by Kosuna (Chapter 8) and Fujita (Chapter 10) and some other scientists, Perilla leaf extracts contain a lot of active constituents showing antioxidative activity such as flavonoids, anthocyanins, phenolics (rosmarinic acid, caffeic acid, caffeates, protocatechuic aldehyde etc.). Venge (1995) reported that oxygen radicals may also play an important role in allergy. Studies indicated that the cells involved in the allergic inflammation are potent producers of various oxygen metabolites. These cells produced large amounts of oxygen radicals which related to the allergic symptoms. From this point of view, it could be suggested that some antioxidants present in Perilla leaves may be involved in its action in the treatment of allergy.

Perilla Essential Oil

Perilla seed, leaf and stem contain a total amount of essential oil about 0.5%. In addition to perillaldehyde, which was removed from the Perilla leaf extract products for its potential allergen property (Kosuna, Chapter 8), several other constituents contained in Perilla essential oil showed pharmacological activity. It was reported (Ke, 1982) that in animal experiments, one of the constituent in the essential oil, β-caryophyllene, showed relaxing action to the windpipe of guinea pig. Also it showed significantly suppressing action to citric acid or acrylaldehyde induced cough. It may partially explain the action of Perilla on anticough and antiasthma. Another constituent, *l*-menthol showed antiitching action thus making Perilla helpful in the treatment of some allergic skin diseases (Yin and Guo, 1994).

Martin *et al.* (1993) have proved the anti-inflammatory activity of α-pinene and β-caryophyllene in the rat edema model induced by carrageenin or by PGE_1. Hashimoto and Fujita (1994) reported that elemicin was found to exhibit an inhibitory action on 5-lipoxygenase (5-LOX) which is the first enzyme in the conversion of arachidonic acid to leukotriene. Elemicin also had a suppressing action in the rat homologous PCA (passive cutaneous anaphylaxis) test which is the most commonly used bioassay to evaluate anti-allergic effects on type I allergy.

Perilla Seed Oil

Earlier, it was proposed that the n-6 and n-3 polyunsaturated fatty acids in diets were importantly involved in allergy (Hashimoto *et al.*, 1988). Recent studies have indicated that the dietary α-linolenate/linoleate balance affected on the development of some chronic diseases including allergy and that raising the (n-3) to (n-6) ratios of diets would be effective in reducing the severity of immediate-type allergic hypersensitivity (Okuyama, 1992; Watanabe *et al.*, 1994). It was hypothesized that if taking more n-3 fatty acid, resulted in a lowering of n-6 fatty acid uptake, this would lead to the inhibition of leukotriene production which is another important mediator causing allergy (Figure 1). Therefore taking Perilla oil (contains mainly n-3 polyunsaturated fatty acid) might be beneficial for the prevention and treatment of allergy.

Ito *et al.* (1992) reported a pilot study on atopic dermatitis patient treated with Perilla oil. In his study, patients treated with Perilla oil, in the phospholipid fraction in serum, the n-3:n-6 ratio and the EPA:AA ratio were significantly increased. Leukotriene C4 release from polymorphonuclea leukocytes by zymosan and fresh autologous serum was significantly decreased. This results might support the hypothesis that increasing α-linolenic acid could inhibit release of some chemical mediators (leukotrienes) linked with allergy.

CONCLUSION

Allergy has become a serious problem affecting health and social economy. New approaches to the treatment of allergy have been studied and developed. Based on the long–term use as a traditional Chinese herbal medicine, food spice and garnish; Perilla has now been confirmed in limited trials to have beneficial effects in the treatment of some allergy diseases such as hay fever and atopic dermatitis in children and adults, apparently without side effects. Preliminary mechanism studies support the use of Perilla products in the treatment of allergy and these are promising candidates for use in clinical double-blind, placebo-controlled trials.

ACKNOWLEDGEMENT

We are thankful to Mr. Arno Latvus, Managing Director of Hankintatukku Natural Products Co., for his considerable support for this work.

REFERENCES

Aaltonen, M. and Kostiainen, E. (1995) Allergy–A recent campaign in Finnish community pharmacies. *Abstracts Book, FIP'95 Stockholm*, No. 283.

Bjorneboe, A., Soyland, E. and Bjorneboe, G.E. (1989) Effect of n-3 fatty acid supplement to patients with atopic dermatitis. *J. Intern. Med. Suppl.* **225**, 233-236.

Bousquet, J., Dhivert, H. and Michel, F-B. (1994) Current trends in the management of allergic diseases. *Allergy*, **49**, 31–36.

Broide, D. (1995) Clinical studies with cetirizine in allergic rhinitis and chronic urticaria. *Allergy*, **50**, 31–35.

Burks, A. W. (1995) The human body and the different reactions to food that may occur. *Allergy*, **50** (suppl. 20), 6–7.

Chen Ke-ji and Chen Kai (1989) Achievements in clinical research of treating internal diseases with traditional Chinese medicine in recent years. *Chinese Medical J.*, **102**, 735-739.

Cooper, K.D. (1994) Atopic dermatitis: recent trends in pathogenesis and therapy. *J. of Investigative Dermatology*, **102**, 128–137.

Emberlin, J. (1994) The effects of patterns in climate and pollen abundance on allergy. *Allergy*, **49**, 15–20.

Haranaka, K., Satomi, N., Sakurai, A, Haranaka, R., Okuda, N. and Kobayashi, M. (1985) Antitumor activities and tumor necrosis factor productivity of traditional Chinese medicine and crude drugs. *Cancer Immun. Immunother.*, **20**, 1–5.

Hardman, R. (1993) New approaches in Chinese traditional medicine and natural drug development. *International Pharmacy J.*, 1993, **7**, 247–249.

Harper, J.I., Yang, S-L., Evans, A.T., Evans, F.J. and Phillipson, J.D. (1990) Chinese herbs for eczema. *Lancet*, **335**, 795.

Hashimoto, A., Katagiri, M., Torii, S., Dainka, J., Ichikawa, A. and Okuyama, H. (1988) Effect of the dietary α-linolenate/linoleate balance on leukotriene production and histamine release in rats. *Prostaglndins*, **36**, 3–17.

Hashimoto, K. and Fujita, T. (1994) Studies on anti-allergic components in the roots of Asiasarum sieboldi. *Planta Med.*, **60**, 124–127.

HayGlass, K.T. (1995) Allergy: who, why and what to do about it? *Immunology Today*, **16**, 505–507.

Imaoka, K., Inouye, S., Takahashi, T. and Kojima, Y. (1993) Effects of *Perilla frutescens* extract on anti-DNP IgE antibody production in mice. *Arerugi*, **42**, 74–80 (in Japanese).

Inaba, M. (1992) Stopping allergic symptoms by orally using Perilla extract. *My Health*, **1992**, No.9. 161–162 (in Japanese).

International Rhinitis Management Working Group (1994a) Mechanisms of rhinitis. *Allergy*, **49** (suppl. 19) 7–9.

International Rhinitis Management Working Group (1994b) Treatment of allergic rhinitis. *Allergy*, **49** (suppl. 19) 19–23.

Ito, K., Kikuchi, S., Yamada, M., Torii, S. and Katagiri, M. (1992) Effect of the α-linolenic acid enriched diet on atopic dermatitis. A pilot study on 6 outpatients. *Jap. J. of Pediatric Allergy and Clinical Immunology*, **6**, 87–91.

Jiangsu New Medical College (ed.) (1977) *Dictionary of Chinese Materia Medica* Vol. **2**, Shanghai Science-Technology Publisher, p. 2357 (in Chinese).

Kabaya, S. (1994) Perilla extract used for atopic dermatitis. *Sawayaka Genki*, **1994**, 196–202 (in Japanese).

Kawakita, T., Nakai, S., Kumazawa, Y., Miura, O., Yumioka, E. and Nomoto, K. (1990) Induction of interferon after administration of TCM, xiao-chai-hu-tang (shosaikoto). *Int. J. Immunopharmac.*, **12**, 515–521.

Kay, B. (1993) Grafting a fresh cure for asthma. *New Scientist*, **1993**, February, 38–42.

Ke Ming qing (ed.)(1982) *Physico-chemical and pharmacological studies on the active compounds in Chinese medicines*, p.74 (in Chinese).

Kosuna, K. (1993) Perilla leaf extract as antiallergy food. *New Food Industry*, **34**, 30–32 (in Japanese).

Kosuna, K. (1995) Studies on anti-inflammatory constituents in Perilla extracts. *Fragrance J.*, **1995**, 7, 90–94 (in Japanese).

Kozo, N. (1994) Self made Perilla leaf decoction for urticaria and pollinosis. *Anshin*, **1994**, 7, 266–267.

Latchman, Y., Bungy, G.A., Atherton, D.J., Rustin, M.H.A. and Brostoff, J. (1995) Efficacy of traditional Chinese herbal therapy *in vitro*. A model system for atopic eczema: inhibition of CD23 expression on blood monocytes. *Brit. J. Dermatol.*, **132**, 592–598.

Leung, R. (1993) Prevalence of allergy and atopy in Hong Kong–a review. *J. of Hong Kong Medical Association*, **45**, 232–238.

Lichtenstein, L.M. (1993) Allergy and the immune system. *Scientific American*, 1993, September, 85–93.

Malling, H.J. (1994) Immunotherapy in Europe. *Clinical and Experimental Allergy*, **24**, 515–521.

Martin, S., Pardilla, E., Ocete, M.A., Galvez, J., Jimenez, J. and Zarzuelo, A. (1993) Anti-inflammatory activity of the essential oil of Bupleurum fruticescens. *Planta Med.* **59**, 533–536.

Mitsuki, S. Experience in the application of Perilla products for allergy. *Anshin*, **1992**, No.7, 175, 177, 179, 181, 185 (in Japanese).

Mizumoto, S. (1992) Perilla extract for rhinitis and atopic dermatitis. *My Health*, 1992, No.9. 159–160 (in Japanese).

Nuutinen, J. (1995) Self medication of allergic rhinitis.*Terveydeksi*, 1995, 8–11 (in Finnish).

Okuyama, H. (1992) *Effects of dietary essential fatty acid balance on behavior and chronic diseases. Polyunsaturated Fatty Acids in Human Nutrition*, eds. U. Bracco and R.J. Deckelbaum, 169–178.

Ortolani, C. and Vighi, G. (1995) Definition of adverse reactions to food. *Allergy*, **50** (suppl. 20), 8–13.

Oyanagi, K. (1993) Experience in treating allergy with Perilla products. *Anshin*, **1993**, No.7, 147, 150, 152, 155.

Puls, K.E. and Bock, K.H. (1993) Pollen flight forecasting in Germany and in Europe. *Experientia*, **49**, 943–946.

Rubenstein H. S. and Rubenstein, J.S. (1984) Impressions of clinical allergy in China. *JAMA*, **252**, 3127.

Rusznak,C., Devalia, J.L. and Davies, R.J. (1994) The impact of pollution on allergic disease. *Allergy*, **49**, 21–27.

Sheehan, M.P., Rustin, M.H.A., Atherton, D.J., Buckley, C., Harris, D.J., Brostoff, J., Ostlere, L. and Dawson, A. (1992) Efficacy of traditional Chinese herbal therapy in adult atopic dermatitis. *Lancet*, **340**, 13–17.

Sheehan, M.P. and Atherton, D.J. (1994) One-year follow up of children treated with Chinese medicinal herbs for atopic eczema. *British J. of Dermatology*, **130**, 488–493.

Sumimoto, S., Kawai, M., Kasajima, Y. and Hamamoto, T. (1992) Increased plasma TNF-alpha concentration in atopic dermatitis. *Arch. Dis. Child.* **67**, 277–279.

Sutherland, K. (1994) Breathing easy in spring. *The Australian J.of Pharmacy*, **75**, 724–735.

Syvänen, P. (1995) What happens when children get allergy. *Terveydeksi*,**1995**, 12-16 (in Finnish).

Tajiri, N. (1993) Efficacy of Perilla extract in allergic symptoms. *My Health*, **1993**, No.3 198–199 (in Japanese).

Takiguchi, T. (1993) Simply and safely using Perilla extract for effectively improving atopic dermatitis symptoms. *My Health*, **1994**, No.8, 150–151 (in Japanese).

Toda, S., Kimura, M., Ohnishi, M. and Nakashima, K., (1988) Effects of the Chinese herbal medicine "SAIBOKUTO" on histamine release from and degranulation of mouse peritoneal mast cells induced by compound 48/80. *J. of Ethnopharmacology*, **24**, 303–309.

Ullenius, B., Davachi, L., Ericsson, L., Johansson, I.M., Nilsson, A., Nilsson, JLG, Skoglund, A. and Söderlund, L.Å. (1995) The allergy year 95–Improved quality of life for people suffering from allergies. *Abstracts Book, FIP'95 Stockholm*, No. 443.

Umesato, Y., Iikura, Y. and Nakamura, Y. (1984) Asthmatic children and Chinese medicine, scientific study of Shoseiryuto and Saibokuto. *J. Allergology*, **33**, 1047–1052.

Venge, Per. (1995) Do oxygen radicals play a role in asthma and allergy? *Abstracts of 6th International Symposium on Trends in Biomedicine in Finland: Antioxidants, Fatty acids, Trace elements and vitamins in Human Health*, p.14–15.

Vercelli, D. and Geha, R.S. (1989) The IgE system. *Ann. Allergy*, **63**, 4–11.

Watanabe, S., Sakai, N., Yasui, Y., Kimura, Y., Kobayashi, T., Mizutani, T. and Okuyama, H. (1994) A high alpha-linolenate diet suppresses antigen-induced immunoglobulin E response and anaphylactic shock in mice. *J. of Nutrition*, **124**, 1566–1573.

Xu, Hui guang (ed.)(1983) *Knowledge in popular traditional Chinese medicines*. p. 22. Shanghai Scientific and Technology Publisher. (in Chinese).

Yamagata, M. (1992) Evaluation of Perilla extract for the atopic and rhinitis. *Anshin*, **1992**, No.7, 172–173.

Yamazaki, M. (1992) Inhibition by Perilla juice of tumor necrosis factor production. *Biosci. Biotech. Biochem.* **56**, 149, 152.

Yamazaki, M. (1993) Anti-inflammatory and antiallergic activities of Perilla juice. *Fragrance J.* **1993**, No. 9, 75–81 (in Japanese).

Yamazaki, M. (1994) Experimental evidence for the improvement of defense capacity to allergy by Perilla leaf. *Wakasa* 1994, No.4 p.84–85 (in Japanese).

Yin Jian and Guo Li-gong (eds.) (1994) *Modern Research and Clinical Application of Chinese Medicines.* p. 632–633 (in Chinese).

Yu He-Ci (1996) Perilla and antiallergy. Digest of Health and Natural Medicine No. 68 (Internal Publications of Hankintatukku Natural Products Co., Finland).

Zhong , N.S. (1994) Atopic diseases in Chinese community. *Clinical and Experimental Allergy,* **24**, 297–298.

Zhu You-Ping and Woerdenbag, H.J. (1995) Traditional Chinese Herbal Medicine. *Pharm.World Sci.,* **17**, 103–112.

7. A CLINICAL INVESTIGATION OF PERILLA EXTRACT CREAM FOR ATOPIC DERMATITIS

KAZUHIKO OYANAGI

Odori Children's Clinic, Medical Building 3M, Odori Nishi 16-1, Sapporo, 060 Japan

INTRODUCTION

Atopic Dermatitis

Atopic dermatitis is one kind of allergic disease. Allergies are very closely associated with an immune response. When the human body is invaded by a foreign substance (antigen), antibodies or sensitised lymphocytes will be produced as a result of the response of the immune system. Later when the same antigen invades the body again, it will soon be eliminated or become harmless to the body. This is an immune response which is an indispensable function to prevent infection and tumours. However, sometimes the immune reaction between antigen and antibodies or sensitised lymphocytes can cause harm to the body itself. This kind of immune reaction in which antigen comes from outside the body causes allergic disease, whereas antigen which comes from the body itself causes auto-immune disease.

According to the statistical investigation in 1992 by the Ministry of Welfare of Japan, 34% of the Japanese population suffer from some kind of allergy, and most of them are children between the age of 0 to 4. There is the tendency for allergic symptoms to appear as atopic dermatitis in childhood and to become asthma or rhinitis as they mature.

The word atopy is derived from Greek (Okabe, 1990) and means odd and thus atopic dermatitis is a disease unknown in its mechanism and its predisposition. It involves both the over-production of IgE antibody in reaction to environmental antigen as well as hereditary factors.

There are various criteria in the diagnosis of atopic dermatitis. In general, skin diseases with inveterate, chronic, and recurrent symptoms such as itchy, dry are diagnosed as atopic dermatitis or eczema. Atopic dermatitis was thought to be chronic eczema mainly occurring in infants or children. However, recently there has been a remarkable increase in the number of patients of all ages and the body takes a long time to recover. It is unknown why this is so, but it can be thought to be related to the living conditions, change in diet, air pollution, and contamination of water and food by chemical substances.

In Japan atopic dermatitis is becoming a serious problem; adult patients have increased in numbers and the disease has caused social problems such as discrimination at work and unsympathetic treatment by the family. Many methods are used to treat allergies: elimination of allergen, application of steroids and antihistamine, traditional Chinese medicine, folk remedies (mugwort, peach leaves, houttuynia), ultraviolet treatment as well as psychotherapy (Okabe, 1990). However, some cases of dystrophia are caused by the elimination of allergen from the diet and adverse reactions often occur because of such medicine.

Background to How Perilla Extract came to be used for Atopic Dermatitis

Many cases of atopic dermatitis require considerable time and effort. The cause of the disease is complex, and the range of the patients' age bracket is widening, therefore, there are various remedies being used. For curative medicine, antihistamines and steroids are often used and have been shown to be effective in many cases. On the other hand, side effects caused by perocutaneous absorption may decrease the effectivity of the medicine because of the repetitive use (Okuhira, 1993). More and more people have begun to worry and now refuse to use steroids since there has been so much exaggerated information given by the media. Psychological aspects, such as stress, are one of the primary causes of atopic dermatitis. Therefore, the patients' anxieties regarding the use of antihistamines and steroids cannot be ignored. This is why various treatments using naturally occurring substances have been examined.

Perilla extract cream was chosen for atopic dermatitis because of the reports from Dr. Yamazaki, at Teikyo University, that Perilla extract had an anti-inflammatory value inhibiting the production of Tumour Necrosis Factor (TNF). Perilla is eaten as a vegetable and in spices in Japan, and the patients can take it without any subsequent problem.

Perilla leaves contain perillaldehyde, an irritating substance, which was removed from the Perilla extract in the following clinical test.

CLINICAL TEST USING PERILLA EXTRACT CREAM

In investigations involving the usefulness of Perilla extract cream for the skin, it was shown to keep skin moist and to prevent dryness. Perilla extract is a natural substance and Perilla extract cream is not an ointment which is regarded as "medicine". Active surface agents were used as little as possible and any ingredient or compound considered to be an irritant was removed or omitted.

Test Method

The patients

The patients with atopic dermatitis who had agreed to this clinical test were examined. Those with an advanced disease were judged to be inappropriate for this study. Over 90 patients participated. Most of them were children under five years old. The sex, age and background of the patients are given in Table 1.

The tests were performed using Perilla extract cream which contained 1% Perilla extract (Group A) or 5% extract (Group B) and later 3% extract (Group C) was used (Table 1).

Among the patients, about 26% of them also suffered from other complications such as allergic rhinitis, bronchial asthma, wheeze, hypophyseal dwarfism, and histidinemia.

Table 1 Background to patient symptoms (95 Cases)

Observation Items		No of Cases (%)		
		Group A	Group B	Group C
sex	male	14 (41.2%)	15 (50.0%)	19 (61.3%)
	female	20 (58.8%)	15 (50.0%)	12 (38.7%)
age group (years old)	0	0 (0.0%)	4 (13.3%)	0 (0.0%)
	1–2	15 (44.1%)	9 (30.0%)	10 (32.3%)
	3–5	12 (35.3%)	6 (20.0%)	15 (48.4%)
	>6	6 (17.7%)	11 (36.7%)	6 (19.3%)
	unknown	1 (2.9%)	0 (0.0%)	0 (0.0%)
diagnosis	atopic dermatitis	29 (85.3%)	30 (100.0%)	30 (96.8%)
	eczema	5 (14.7%)	0 (0.0%)	1 (3.2%)
severity				
eruption	very severe	4 (11.8%)	3 (10.0%)	5 (16.1%)
	severe	18 (52.9%)	20 (66.7%)	23 (74.2%)
	mild	4 (11.8%)	7 (23.3%)	1 (3.2%)
	slight or nil	0 (0.0%)	0 (0.0%)	0 (0.0%)
	unknown	8 (23.5%)	0 (0.0%)	2 (6.5%)
age symptoms started	0	21 (61.8%)	15 (50.0%)	19 (61.3%)
	1	7 (20.6%)	5 (16.7%)	4 (12.9%)
	2-6	4 (11.7%)	9 (30.0%)	5 (16.1%)
	unknown	2 (5.9%)	1 (3.3%)	3 (9.7%)

Investigation showed that in about 30% of the patients one or more of their family suffered from atopic dermatitis. More than 20% of all the patients complained that the occurrence of the allergy was associated with a change in the weather. More than 24% of the patients reported their allergy was diet related.

Most of the patients used such medicines as steroids, antiallergic agents and antihistamines before coming into this trial.

Those patients in Group B (receiving 5% extract cream) contained a higher percentage of patients with advanced disease than those in Group A (receiving 1% cream).

Side effects were recorded by the degree, symptoms, the relation of cause and effect, and the date of occurrence.

Application Method and Period

Perilla extract was applied two to three times daily onto skin at the affected part and judged every 2 weeks and 5 times in 8 weeks. Data was recorded on the designated cards. The average period of observation was 51.5 days in Group A, 35.1 days in Group B, and 61.4 days in Group C.

KAZUHIKO OYANAGI

Table 2 Improvement of Symptoms

Group A (1% Cream, 32 cases)

Observation Item	Non observed (case)	Observed (case)	Symptoms totally improved	improved	unchanged	aggravated	Improvement %
itching eruption	0	32	3 9.4%	21 65.6%	8 25.0%	0 0.0%	75.0%
erythematous eruption	11	21	6 28.6%	8 38.1%	7 33.3%	0 0.0%	66.7%
papular eruption	19	13	4 30.8%	2 15.4%	7 53.8%	0 0.0%	46.2%
desquamative eruption	6	26	8 30.8%	10 38.4%	8 30.8%	0 0.0%	69.2%
infiltrating eruption	17	15	8 53.3%	1 6.7%	6 40.0%	0 0.0%	60.0%

Group B (5% Cream, 31 cases)

Observation Item	Non observed (case)	Observed (case)	Symptoms totally improved	improved	unchanged	aggravated	Improvement %
itching eruption	0	31	3 9.7%	21 67.8%	6 19.3%	1 3.2%	77.5%
erythematous eruption	0	25	2 8%	10 40.0%	12 48.0%	1 4%	48.0%
papular eruption	10	21	4 19.0%	3 14.3%	13 61.9%	1 4.8%	33.3%
desquamative eruption	0	31	6 19.4%	12 38.7%	11 35.5%	2 6.4%	58.1%
infiltrating eruption	20	11	3 27.3%	2 18.2%	6 54.5%	0 0.0%	45.5%

Group C (3% Cream, 30 cases)

Observation Item	Non observed (case)	Observed (case)	Symptoms totally improved	improved	unchanged	aggravated	Improvement %
itching eruption	0	30	7 23.3%	19 63.3%	3 10.0%	1 3.3%	86.7%
erythematous eruption	0	30	8 26.7%	15 50.0%	6 20.0%	1 3.3%	76.7%
papular eruption	21	9	4 44.4%	2 22.2%	3 33.3%	0 0.0%	66.7%
desquamative eruption	7	23	6 26.1%	11 47.8%	6 26.1%	0 0.0%	73.9%
infiltrating eruption	5	25	8 32.0%	11 44.0%	6 24.0%	0 0.0%	76.0%

Evaluation and Improvement of Symptoms (Table 2)

Symptoms were divided into 5 types: itching eruption, erythematous eruption, papular eruption, desquamative eruption, and infiltrating eruption. They were evaluated at five levels: high degree, middle degree, low degree, slight degree, and none. Two weeks after the treatment with Perilla cream, the above symptoms and also redness were evaluated. The improvement of individual symptom was evaluated at four levels (totally improved, improved, unchanged and aggravated). The general improvement was evaluated at five levels (highly improved, improved, slightly improved, unchanged, and aggravated).

As a whole, effectiveness of Perilla cream was evaluated collectively from the skin symptom, the general improvement, and its side effects after treatment by following four levels: highly effective, effective, slightly effective, and ineffective.

Results of the Treatment

Symptoms improvement

Table 2 shows the improvement in the symptoms in all the three groups. In each group symptom itching eruption was improved very well. In Group A and B symptom papular eruption was less improved and Group C showed the most satisfying improvement for all the symptoms.

General improvement

General improvement using 1% (Group A), 5% (Group B), and 3% (Group C) Perilla cream are shown in Figure 1. The improvement was recognised in each group: 73.5% in Group A, 80.6% in Group B, and 83.4% in Group C.

Effectiveness

In every case, no side effect was found. The effectiveness percentages of Perilla cream for the treatment of atopic dermatitis is shown in Figure 2. The effectiveness were 70.6%, 80.6%, and 80.0% in the three groups after using 1%, 5%, 3% Perilla cream respectively.

Furthermore effectiveness was evaluated according to the patients conditions such as effectiveness whilst using other medicines (Figure 3), effectiveness on pruritus (Figure 4), effectiveness on different types of eruption (Figure 5), and effectiveness after the use of steroids had been discontinued (Figure 6).

For the effectiveness of Perilla cream in the absence of other medicine (Figure 3) there was not much difference between Group A and Group B, 60% and 68% effectiveness, respectively.

When comparing the effectiveness of 1%, 5% and 3% creams, for pruritus (Figure 4) 5% cream had a higher effectiveness than 1% cream but in mild cases 3% cream appeared even better. Concerning the dry type of infantile eczema (Figure 5) both 5% and 3% creams had a high effectiveness but there was insufficient data to say which strength cream was to be preferred and likewise for other types of eruptions.

Group A (34 cases)

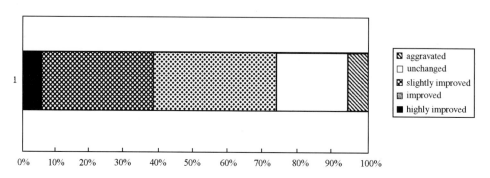

Group B (31 cases)

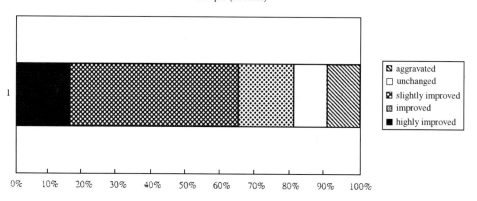

Group C (30 cases)

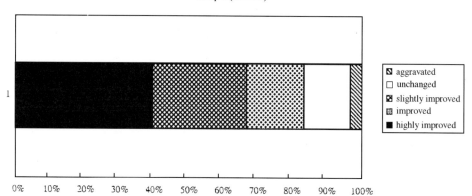

Figure 1   General improvement

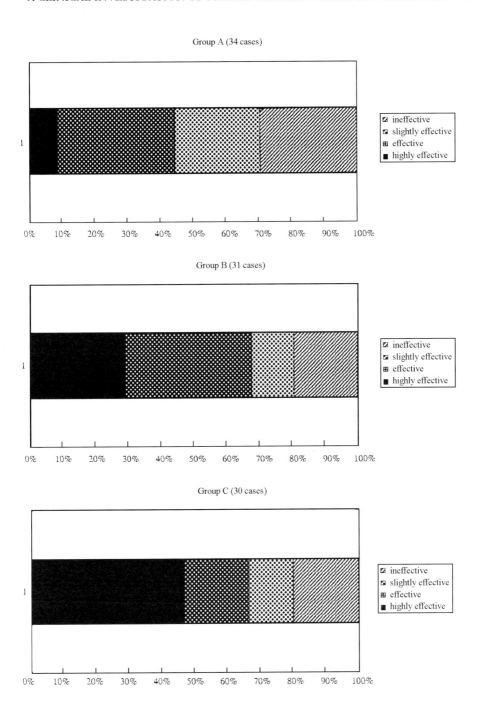

Figure 2   Effectiveness

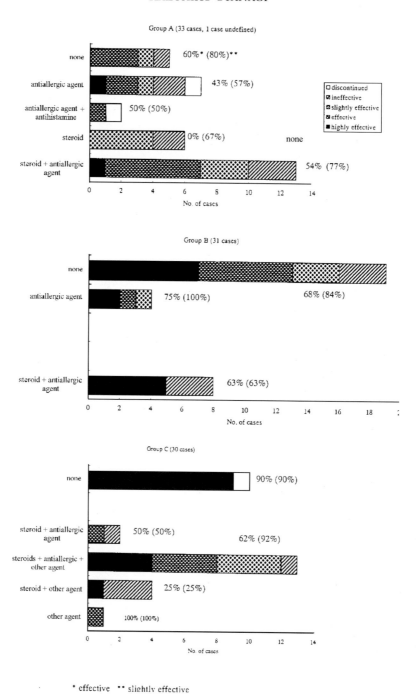

Figure 3   Effectiveness of Perilla cream when used with and without other medicines

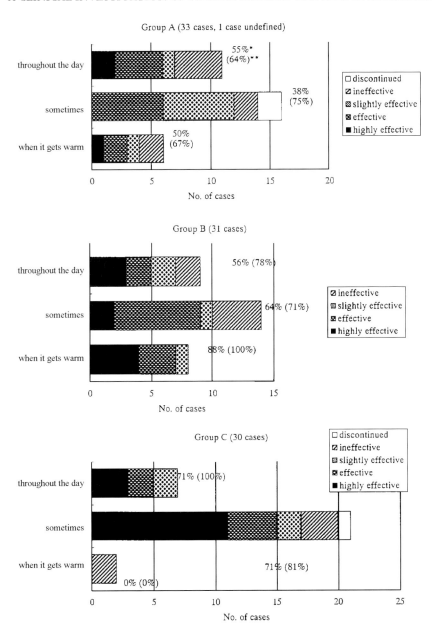

Figure 4   Effectiveness of Perilla cream on pruritus

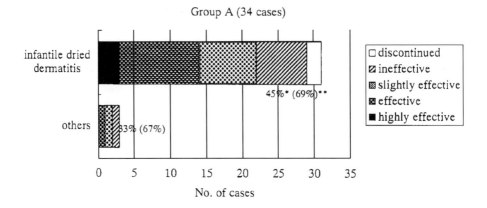

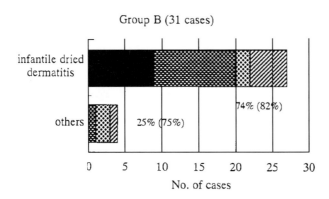

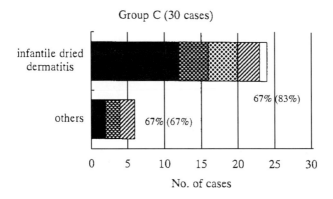

*effective **slightly effective

Figure 5   Effectiveness of Perilla cream on different types of eruption

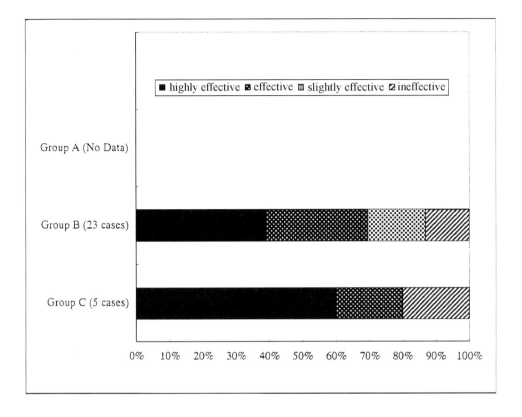

Figure 6   Effectiveness of Perilla cream (after steroid had been discontinued)

Steroid is most effective as an ointment for application, however, it has side effects. A separate preliminary study was conducted with 5% and 3% Perilla extract cream in the absence of steroids (Figure 6). In a comparison of the patients who stopped using steroids, the patients using anti-allergic medicine for external and also internal use were included in this examination. Because of the difference in the number of cases, it is difficult to compare the two groups. However, Group B has a very high effectiveness and a decrease in the amount of steroids used is to be expected based on the results of this preliminary study.

OBSERVATION SUMMARY

The percentage of the improvement in patients in Group A was 73.5% and in Group B, 80.6%. Group C showed improvement equivalent to that of Group B with the test which was given later under the same conditions. From these results, even a small amount of Perilla extract helps to improve patients with atopic dermatitis.

In the case of unchanged and aggravated results, the reason was not specified as there was no common case. In most improved cases, symptoms became aggravated after 4 to 5 days, whereafter continued application improved them within two weeks. It is presumed to be a kind of rebound phenomenon. More than 70% of the patients requested to continue the application even after the study ended and their comments on using the cream were quite favourable.

Concerning the efficacy of Perilla cream, of the 90 patients, 70–80% experienced some kind of improvement. On the whole, among the extracted natural substances, Perilla extract was recognised to be highly beneficial for atopic dermatitis. Considering there was no side effect in any of the cases, perillaldehyde free Perilla extract is expected to be used as anti-allergy medicine for a therapy in the future, and more extended trials are justified.

REFERENCES

Okabe, S. (1990) *Therapy of Traditional Chinese Medicines for Atopic Dermatitis*, Gendai Shuppan Planning, 14–23 (in Japanese).

Okuhira, H. (1993) Medical treatment of atopic dermatitis in place of steroids. *Nikkei Science*, Nikkei Shinbunnsha, 6–10 (in Japanese).

8. THE DEVELOPMENT AND APPLICATION OF PERILLA EXTRACT AS AN ANTI-ALLERGIC SUBSTANCE

KENICHI KOSUNA and MEGUMI HAGA

Amino Up Chemical Co., Ltd., High Tech Hill Shin-ei, 363-32 Shine-ei, Sapporo, Japan 004

INTRODUCTION

In our laboratory we have been seeking out and researching natural substances contained in plants and food which are beneficial to the human body. Perilla extract was discovered and developed partially because of this research.

Recently there has been a marked increase in the number of people who have allergies. To find effective treatments for allergies, a great effort has been made in the medical field. However, a satisfactory treatment has yet to be established. It is particularly difficult because most allergies are caused by living habits and conditions, environmental factors, eating habits, individual predispositions including genetic causes, and stress related factors. The treatment is often lengthy.

In order to prevent allergies, possible allergens have been removed from our diet. However, this has resulted in an unbalanced diet and diseases which have been caused by a loss of tolerance to allergens, especially in children. For treatment and prevention, we thought of using those physiologically active substances contained in foods which would be beneficial to the human body. Therefore, they should be harmless in large amounts and familiar to the diet. We started by analysing the effects of various vegetables on the human immune system and observed the reactions. As a result of this research, Perilla extract was found to be an effective anti-allergy substance (Oyanagi *et al.*, 1992).

SCREENING OF FOODS FOR ANTI-ALLERGY PROPERTIES

The effects of food on biophylaxis, that is the immune system of the body, is now receiving much attention. Cytokines, a bacterial formulation of BRM (biological response modifier) and certain foods (vegetables, mushrooms, fruit) boost the immune system and enhance the body's self-defence system (Ueda *et al.* 1991a, b; Yamazaki, 1992; Yamazaki and Nishimura 1992) whereas some steroids and other foods inhibit the self-defence system. The self-defence system needs to be regulated according to the body's condition. BRMs and steroids are used in order to intentionally control it. BRMs are used for enhancement of cellular immunity in cancer treatment. Steroids are used as anti-allergy agents. With this consideration, we have focused on the fact that food and food substances have an immunoregulatory influence.

Dr. M. Yamazaki of Teikyo University in Japan, adopted the TNF (Tumour Necrosis Factor) which is produced by the leukocyte, as a measure of the activity of the body's

Table 1 Neutrophil accumulation by vegetable juice

Sample	% of Neutrophil	ED_{50} (ml/mouse)
Control	2 ± 1	-
Turnip	42 ± 7	1.4
Japanese radish	45 ± 8	0.7
Cucumber	51 ± 5	0.6
Green pepper	62 ± 9	0.4
Eggplant	59 ± 3	0.4
Parsley	57 ± 7	0.3
Carrot	71 ± 8	0.2
Spinach	94 ± 3	0.18
Spring onion	78 ± 5	0.15
Cabbage	74 ± 8	0.09
Ginger	76 ± 4	0.08
Onion	82 ± 7	0.07
Perilla	89 ± 6	0.02
Garlic	72 ± 7	0.02

Control : Saline
ED_{50}: the effective dose when Neutrophil accounts for 50% of the all cells

Table 2 Nutritional value of green and purple Perilla leaves (100 g)

Moisture		87.5 g
Protein		3.8 g
Lipid		0.1 g
Carbohydrate	Carbohydrate	5.5 g
	Fibre	1.5 g
Mineral	Calcium	220 mg
	Phosphorous	65 mg
	Iron	1.6 mg
	Sodium	1 mg
Ash		1.6 g
Vitamin	A Retinol	0 μg
	Carotene	8700* μg
	A-effect	4800 IU
	B1	0.12 mg
	B2	0.32 mg
	Niacin	1.0 mg
	C	55 mg

*Purple Perilla 7200

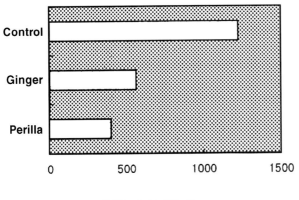

TNF activity (U/ml)

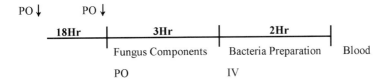

U/ml : Unit per ml (1 Unit is the 50% concentration that TNF affects L929 cells)
PO : per os
IV : intravenous

Figure 1 Inhibition of TNF production by food.

self-defence system (Ueda *et al.*, 1991a,b). We have focused on foods which inhibit excessive TNF production so as to reduce the adverse effects of allergies.

For screening for BRM, accumulation of blood neutrophil and TNF production activity were used and Perilla's inhibition of TNF production was found from this screening (Yamazaki *et al.*, 1992; Yamazaki and Ueda, 1995). Neutrophil accumulation of various vegetable components was examined and some of them showed high activity (Table 1). Perilla and ginger showed a higher TNF inhibiting activity as it showed in Figure 1 (Yamazaki, 1992). Dr. Yamazaki and our laboratory staff have been collaborating in discovering food substances which regulate the functioning of the immune system and Perilla was chosen for further studies.

Green and purple Perilla leaves have been in the diet of the Japanese people for a long time. The nutritional value of the leaves is given in Table 2. However, the leaves are ingested in small amount due to their unique odour and flavour which is mainly caused by perillaldehyde in essential oil of the leaves. The development of a Perilla extract has been focused on obtaining one with anti-allergic effects and the perillaldehyde has been excluded.

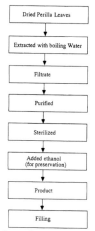

Figure 2   Manufacturing process of anti-allergic Perilla extract

MANUFACTURING PROCESS FOR PERILLA PRODUCTS

Material

The material used was limited to Perilla leaf from Hokkaido (Japan) from the view point of its quality, its anti-allergic effects and the control of its production. The variety used in the manufacturing process was *Perilla frutescens* var. *acuta* Kudo forma *viridis* Makino.

Extraction of Perilla Leaf

The decoction of the dried leaves afforded by boiling water gave an extract (Figure 2) with greater anti-allergy activity than that obtained using aqueous ethanol and avoided the precipitation problems caused by chlorophyll etc., in the aqueous ethanol extract (Kosuna *et al.*, 1994).

Perillaldehyde is considered to be the main cause of contact dermatitis experienced by Perilla farmers (Okazaki and Matsunaga, 1981). During our research, perillaldehyde was proved not to inhibit TNF production as it showed in Figure 3 (Yamazaki, 1994). The extraction and purification process yielded an extract free of perillaldehyde, protein and lipids all of which can be allergens. The resulting extract was a light reddish brown in colour and has a slight characteristic odour.

Spray Drying

By the addition of dextrin to the aqueous Perilla extract, followed by spray drying, a stable powdered form was achieved, namely Perilla extract powders with 1 g equivalent to 1 ml of Perilla extract. This powder was available for capsule and tablet formulations and for use in a range of health products.

	TNF (U/ml)	Inhibition (%)
Perillaldehyde	2047	- 32
Perilla Juice	732	54
Control	1562	0

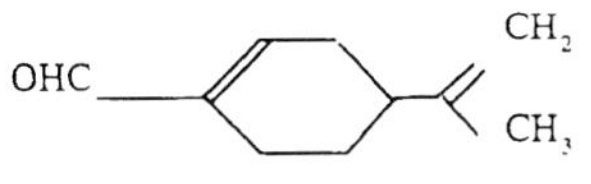

OHC ⎯⎯ $\diagup\!\!\diagdown$ ⎯ $\diagup\!\!\diagup$ $< \begin{matrix} CH_2 \\ CH_3 \end{matrix}$

Figure 3 Inhibition of TNF activity by perillaldehyde.

Application of Perilla Extract to Various Products

A special feature of the aqueous extract was that it did not contain excessive colouring matter, flavour, nor stimulating substance. Hence it could be used in various products for daily use (Kosuna, 1993). For example, it was available for use in foods (drinks, breads, sweets etc.), cosmetics (hair and body shampoo, skin cream, skin lotion, soap etc.), and in medicines (external and internal medicines for allergic dermatitis) (Figure 4).

In recent years, as people consider health care more and more, Perilla extract has been increasingly used in various products. On the Japanese market Perilla is now used in 80 products (including the toiletries and cosmetic range) and in 15 other kinds (including health foods such as encapsulated products) and the range is still growing (March, 1995).

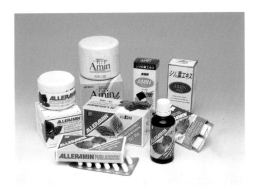

Figure 4 Some products developed from Perilla extract.

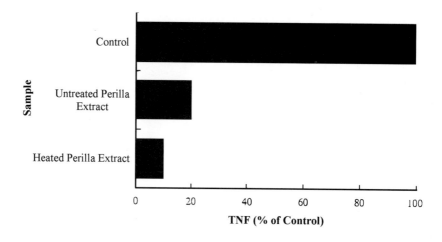

Control: Saline

Figure 5   Heat resistance of Perilla extract.

For the product of Perilla extract it was highly concentrated and remained sterile when in a small bottle. It was effective in a small dose and it was not injurious to health even if taken in excess (Kosuna, 1992). For practical use, 0.4 to 0.5 ml for adults and 0.1 to 0.3 ml for children could be added to water, juice, milk, tea, or soup. The effectivity did not change after adding to hot drinks.

INVESTIGATION OF PERILLA EXTRACT

Product Specification

Analysis showed that Perilla extract contained 0.4% of carbohydrates, 0.1% of ash, and 99.4% of water. Only 0.01 mg of vitamin B_2 was present in 100 g. Formaldehyde, arsenic, lead, and residue of organochlorine pesticides were not detected. As microelements, 6 mg of phosphorus, and 43 mg of sodium were present in 100 g of Perilla extract.

Factors Affecting Product

Heat Resistance Examination

Perilla extract was heated for 10 minutes at 100° C. The TNF inhibiting activity of the heated sample was similar to that of the unheated one (Kosuna, 1992; Yamazaki, 1992). The active component has not been identified yet. However, it is presumed that the active component is contained in a saccharide which is stable for the time and temperature used as it showed in Figure 5 (Kosuna, 1993; Ueda and Yamazaki, 1993).

Effects of pH

When using Perilla extract in products, there was concern that the pH (hydrogen ion) of the mixture would have some influence on the Perilla extract. The influence of pH on the Perilla extract was studied. One month after adjustment of pH there was no precipitation at pH 5.0 to 11.0 but precipitation occurred at other values. The colour darkened between pH 2.0 and pH 12.0.

Effect of Ion

These salts, NaCl, NH_4Cl and Na_2SO_4, when added to the Perilla extract caused no precipitation whereas precipitation occurred with $CaCl_2$ and $FeCl_2$ and with a high concentration of Na_2CO_3 and Na_2HPO_4.

Effect of Surfactants

When these surfactants were mixed with Perilla extract no precipitation occurred: sodium lauroylmethylalanine, 2-alkyl-N-carboxymethyl-N-hydroxyethyl imidazolynium betaine, polyoxyethylene-20 sorbitanmonostearate, polyoxyethylene-20 hydrogenatedcastoroil. Precipitation occurred with benzalkonium chloride.

Effects of Solvent

Solvents such as ethanol and isopropanol when mixed with the Perilla extract caused precipitation whereas propylene glycol and glycerol did not do so.

Preservation Stability Test

Perilla extract was obtained by using hot boiling water as mentioned previously. Heat sterilised bottles were filled aseptically. For use as an ingredient, 10% of ethanol was added as a preservative agent. Chemical additives were not used for disinfection nor stabilisation. For confirmation of stability and sterilisation, the variation of colour (O.D. at 400 nm) and contamination by bacteria were checked. Storage was carried out for 8 months. There was no variation in colour at 5°C but the colour darkened at 25°C and 37°C. After 8 months the fructose spot test showed the absence of bacteria.

Toxicity and Safety

Safety of Perilla extract was examined.

Acute Toxicity Test

At Teikyo University the Perilla extract was tested for acute toxicity by using 8 weeks old C3H/He inbred mice from Shizuoka Experimental Animal Farm (10/group) and intraperitoneal dosage of 10.0 ml, 5.0 ml, 1.0 ml, and 0.2 ml. The activity of the 10 ml

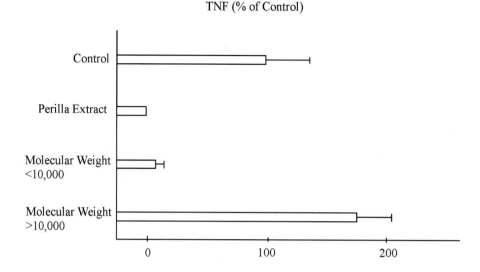

TNF (% of Control)

Control : Saline

Figure 6 Inhibition of TNF production (with demarcation through ultrafilter membrane)

group slowed down for a few hours after receiving the Perilla extract, recovering to normal on the next day. Observation was continued for one month and there was no weight loss or death.

Influence of Continuous Administration

Also at Teikyo University two groups of 8 week old C3H/He inbred mice were allowed to take tap water containing 1.0% and 0.1% Perilla extract, respectively, freely for 31 days. The mice showed no aversion to this water. There was no difference in weight variation compared with a control group. The amount of feed taken was virtually the same also. The body condition and behaviour of the two test groups were normal and no different from the control group (Kosuna, 1992; Kosuna, 1994).

The internal organs, liver, spleen, kidney and thymus are closely associated with the immune system, such as the production of macrophages, B cells, T cells and cytokines. After the administration of the Perilla extract there was no difference in the weight and appearance (before and after dissection) of these organs when compared with those from the control group. The administered amount used for mice (1.0% group) is equivalent to approximately 100 ml of Perilla extract given to a human. However, it is considered that this high dosage would have no adverse influence on the normal functions of the immune and metabolic system.

Table 3 Inhibition of TNF production of components in Perilla extract

Sample	ED_{50} (mg/ml, mean $\pm$ SD)
Perilla Extract	0.42 ± 0.27
Protocatechualdehyde	0.11 ± 0.04
Caffeic acid	0.13 ± 0.05
Rosmarinic acid	0.13 ± 0.04

ED_{50} : the effective dose of 50% inhibition of control TNF production from Macrophage by different compounds

The Active Principles

The inhibition of TNF production was measured using mice. Perilla extract was administered simultaneously with a priming agent (fungus components) normally giving rise to intense inflammation, TNF production was inhibited by 70–80%. It was fully inhibited (99.5%) by steroids. It can be said that it is preferable not to inhibit TNF production totally, because TNF is an important substance for the body's defence system.

In order to detect key compounds with anti-allergy properties, inhibition of TNF production was examined (Kosuna et al., 1995) with selective demarcation through an ultrafilter membrane with a cut-off of molecular weight 10,000 (Figure 6). Perilla extract was found to be effective for mice endoabdominal macrophage cultured in vitro. The examination results were similar to that of in vivo (Yamazaki, 1995). Thus, this in vitro examination was adopted because it was economical and simple. As a result, inhibiting TNF production was confirmed with material of molecular weight under 10,000.

So far, by chromatographic separation etc., ten fractions from the Perilla extract have been shown to inhibit TNF production. The main compounds identified have been protocatechualdehyde, rosmarinic acid, and caffeic acid. However, the natural plants contain a lot of compounds which are in a tiny amount and difficult to be isolated. It could be thought that some other compounds altogether inhibit TNF production with these three compounds. Their inhibition of TNF production is shown in Table 3, where the ED_{50} is the effective dose of 50% inhibition of control TNF production from macrophage by different compounds (Kosuna et al., 1995).

SUMMARY

Perilla extract inhibits TNF production and has value as an anti-allergy agent. The extract is now used in various products such as foods, cosmetics, and other items and for these the demand is increasing. In a preliminary clinical trial, a perillaldehyde-free cream applied to the skin has been shown to be beneficial in the treatment of atopic dermatitis. Research is continuing on the anti-allergy principles of the extract and it is hoped to develop a more active extract in the future.

REFERENCES

Kosaka, H., Ueda, H., Mano, T., Kunii, H., Onami, M. and Yamazaki, M. (1994) Inhibition of TNF production from macrophage by Perilla juice. *Japanese Society for Bioscience, Bioscience, Biotechnology and Agrochemistry*, **1994,** 300 (in Japanese).

Kosuna, K. (1992) The application of Perilla leaf extract as anti-allergy food substance. *New Food Industry*, **34,** 30–32 (in Japanese).

Kosuna, K. (1993) The development and application of Perilla extract. *Japanese Journal of Food Processing* , **28,** 46–47 (in Japanese).

Kosuna, K. (1994) Self-defence activity by food. *New Food Industry* **36,** 41-44 (in Japanese).

Kosuna, K., Shirai, J. and Kosaka, H. (1995) Anti-inflammatory active component of Perilla extract. *Fragrance Journal,* **1995,** 90–94.

Okazaki, N. and Matsunaga, M. (1981) Dermatitis by Perilla. *Japanese Journal of Skin Disease Treatment,* **3,** 713–716 (in Japanese).

Oyanagi, K., Nihira, M., Tsuchiyama, A., Aoyama, T. and Itakura, Y. (1992) A clinical investigation of Perilla extract cream for atopic dermatitis. *Japanese Society of Pediatric Dermatitis,* **51,** 65 (in Japanese).

Ueda, H., Fukuda, K., Nishimura, T. and Yamazaki, M. (1991a) Activation and inhibition of self-defence system by vegetable juices. *The Pharmaceutical Society of Japan,* **1991,** 162 (in Japanese).

Ueda, H., Okamoto, M., Yui, S. and Yamazaki, M. (1991b) Augmentation and inhibition of TNF release by vegetable juices. *Japanese Society of Immunology,* **1991,** 394 (in Japanese).

Ueda, H., and Yamazaki, M. (1993) Inhibitory activity of Perilla juice for TNF-α production. *Japanese Journal of Inflammation,* **13,** 337–340 (in Japanese).

Yamazaki, M. (1992) *Food and Biophylaxis,* Kodansha Science, Murakami, H. and Kaminogawa, S.(eds.) 136 (in Japanese).

Yamazaki, M. (1994) Host mechanism and immune modulating food. *Research Series,* (Japanese Society of Drug Industry and Health Food Research) **4,** 33 (in Japanese).

Yamazaki, M. (1995) Immunological control activity of TNF-α. *Fragrance Journal,* **5,** 43–48 (in Japanese).

Yamazaki, M. and Nishimura, T. (1992) Induction of neutrophil accumulation by vegetable juice. *Biosci. Biotech. Biochem.,* **56,** 150–151.

Yamazaki, M. and Ueda, H. (1995) Inhibitory activity of Perilla extracts for TNF-α production. *World Congress on Inflammation, The Abstracts of Inflammation Research* , **44,** S278.

Yamazaki, M., Ueda, H. and Du, D.(1992) Inhibition by Perilla juice of Tumour Necrosis Factor production. *Biosci. Biotech. Biochem,* **56,** 152–153.

9. LIPID COMPOSITION AND NUTRITIONAL AND PHYSIOLOGICAL ROLES OF PERILLA SEED AND ITS OIL

HYO-SUN SHIN

Department of Food Science and Technology
Dongguk University, Seoul 100-715, Korea

INTRODUCTION

The ancient traditional use of Perilla seed and its oil as food or medicine in Southeast Asia is recorded in such old books as "Shen Nong Ben Cao Jing" of China, "Dongeuibogam" of Korea and "Engishikiten" of Japan (Chang, 1995).

In Korea roasted Perilla seed is used widely as flavouring and nutritional sources and mixed with other cereals or vegetables. Perilla oil is used as salad oil and cooking medium because of its distinct flavour. The oil is extracted by mechanical pressing of the roasted seed, then purified by a simple refining process such as filtration. The intact leaves of the plant are used as condiments or flavouring agents for several foods in Korea and Japan, and often deep-fried with batter. Koreans have consumed fresh Perilla leaves with grilled red meat for a long time.

Because Perilla oil is highly unsaturated, it has been used mainly for industrial purposes. Its consumption as a food has been limited to a few countries (Sonntag, 1979). People now recognise the dietary significance of α-linolenic acid (Holman *et al.*, 1982; Bjerve *et al.*,1989). Perilla seed and oil are good sources of the α-linolenic acid and this and other aspects of their dietary value are being researched.

LIPID COMPOSITION

Total Lipid Content and Some Physicochemical Characteristics

Perilla seed contains about 38–45% of lipid (Sonntag, 1979; Vaugham, 1970). However, its content varies depending on the variety and growing conditions. Total lipid content of seed of 5 different lines of *Perilla frutescens* Britt. grown in Korea, quantified by the Soxhlet method with ethyl ether, varied from 38.6 to 47.8% (Shin and Kim, 1994). The seed of two varieties of *Perilla* grown in Japan (*P. frutescens* Britt. var. *crispa* Decaisne and *P. frutescens* Britt. var. *acura* Kudo) were extracted with chloroform-methanol (2:1, v/v). The total lipid content was 25.2–25.7% (Tsuyuki *et al.*, 1978). On the other hand, total lipid content of Perilla seed from India (*P. frutescens*) was very high (51.7%), when determined by the Soxhlet method with ethyl ether (Longvah and Deosthale, 1991). These results show that the total lipid content of Perilla seed varies not only with the variety and environmental factors but also with the nature of the extraction method.

Table 1   Major lipid classes of Perilla seed[a]

Cultivar	Neutral lipids		Glycolipids		Phospholipids	
	wt[b]	%[c]	wt[b]	%[c]	wt[b]	%[c]
Suwon 8	36.9	93.5	1.5	3.9	1.1	2.7
Suwon 10	40.7	93.9	1.8	4.2	0.8	2.0
Suwon 21	42.7	93.7	2.0	4.4	0.9	2.0
Suwon 24	35.2	91.2	2.2	5.8	1.2	3.0
Yaebsil	44.2	92.5	2.5	5.2	1.1	2.3

[a]From Shin and Kim, 1994.
[b]Percentages of the seed on dry weight basis.
[c]Percentages represent the fraction of a given lipid class with respect to lipid
content within a cultivar.

The physicochemical characteristics of lipid extracted from Perilla seeds are as following: 1.4760–1.4784 of refractive index (at 25°C), 192.0–196.3 of iodine value (Wijs), 192.7–197.7 of saponification value and 1.3–1.8% of unsaponifiable matter content (Shin and Kim, 1994). Since Perilla seed oil is much higher in iodine value than other vegetable oils, it has the characteristics of a strong drying oil. Thus, Perilla oil is regarded as an inedible oil and instead is used for manufacturing such industrial products as varnishes, printing ink, linoleum, etc. (Sonntag, 1979; Vaugham, 1970).

Lipid Classes

As shown in Table 1, the major lipid classes of Perilla seeds are 91.2–93.9% of neutral lipids (NL), 3.9–5.8% of glycolipids (GL) and 2.0–3.0% of phospholipids (PL). This composition is similar to that of other oilseeds having NL as the major component. NL fraction has 88.1–91.0% of triacylglycerols, 4.1–7.5% of sterol esters and hydrocarbons, 1.9–2.7% of free sterols, and a small amount of free fatty acids and partial glycerides (monoglycerides and diglycerides) (Table 2). Esterified steryl glycoside (48.9–54.3%) is the major component in the GL fraction. Other fractions are sterylglycoside (22.1–25.4%), monogalactosyldiacylglycerol (14.7–18.6%) and digalactosyldiacylglycerol (7.9–9.4%). The PL fraction has 50.4–57.1% of phosphatidylethanolamine, phosphatidylcholine (17.6–20.6%), phosphatidic acid (13.6–19.9%), and a small amount of lysophosphatidylcholine (2.9–4.0%), phosphatidylserine and phosphatidylinositol (4.8–6.6%). It was reported that there were no differences in the individual composition pattern comprising NL, GL and PL as the major lipid classes of Perilla total lipids (Shin and Kim, 1994).

Tsuyuki et al. (1978) reported that the total lipids of seed of two different Perilla lines were composed of triglycerides (79.79–82.46%), sterol esters (1.74–1.81%), free fatty acids (2.46–2.65%), diglycerides (1.58%), sterol (0.72–0.89%), pigments (3.06–4.18%), monoglycerides (0.59–2.19%), complex lipids (2.37–2.91%) and other (3.19–5.83%). From the seed total lipids they also separated 12 spots of complex lipids by TLC and confirmed 6 kinds of lipids: lecithin, lysolecithin, monogalactosyldiglycerides, cerebrosides, phosphatidylethanolamines and phosphatidylserines.

Table 2 Composition of neutral and polar lipids in Perilla seeds[a]

Lipid class	Suwon 8	Suwon 10	Suwon 21	Suwon 24	Yaebsil
Neutral lipid[c]					
SE, HC	4.1[b]	6.2	5.7	7.5	5.6
TG	91.0	89.1	88.7	88.1	88.2
FFA	0.3	0.4	0.7	0.3	0.8
FS	1.9	2.5	2.5	2.6	2.7
DG	0.8	0.3	0.9	0.4	1.2
MG	1.8	1.5	1.4	1.0	1.4
Glycolipid[d]					
SG	25.4	23.3	22.1	23.2	24.4
DGDG	7.9	9.2	8.1	8.0	9.4
ESG	50.8	48.9	54.3	53.2	51.4
MGDG	15.8	18.6	15.4	15.6	14.7
Phospholipid[e]					
LPC	4.0	3.6	3.6	4.0	2.9
PS, PI	4.8	6.1	6.1	6.5	6.6
PC	20.6	17.6	17.9	19.4	20.2
PE	54.2	55.4	57.1	56.6	50.4
PA	16.3	17.4	15.3	13.6	19.9

[a]From Shin and Kim, 1994.
[b]All values are percentages of each lipid class.
[c]SE, sterol esters; HC, hydrocarbons; TG, triacylglycerols; FFA, free fatty acids;
 FS, free sterols; DG, diacylglycerols; MG, monoacylglycerols
[d]SG, sterylglycoside; DGDG, digalactosyldiacylglycerol; ESG, esterified
 sterylglycoside; MGDG, monogalactosyldiacylglycerol.
[e]LPC, lysophosphatidylcholine; PS, phosphatidylserine; PI, phosphatidylinositol; PC, phosphatidylcholine; PE, phosphatidylethanolamine; PA,
phosphatidic acid.

Park et al. (1983) reported the triglyceride composition of total lipids extracted from Perilla seed by using TLC, HPLC, and GLC; the composition of the essential triglyceride of Perilla oil is 68% of ($C_{18:3}$, $C_{18:3}$, $C_{18:3}$), 6.7% of ($C_{18:2}$, $C_{18:3}$, $C_{18:3}$), 5.9% of ($C_{18:1}$, $C_{18:2}$, $C_{18:3}$), 4.3% of ($C_{16:0}$, $C_{18:3}$, $C_{18:3}$), 3.8% of ($C_{18:1}$, $C_{18:2}$, $C_{18:3}$), 3.2% of ($C_{18:1}$, $C_{18:1}$, $C_{18:3}$), 2.0% of ($C_{16:0}$, $C_{18:2}$, $C_{18:3}$), 1.5% of ($C_{18:2}$, $C_{18:2}$, $C_{18:3}$), 1.0% of ($C_{16:0}$, $C_{18:1}$, $C_{18:3}$), etc. The main characteristic is that trilinolenate accounts for 68% of total triglyceride.

Fatty Acid Composition

The major fatty acids of total lipids extracted from Perilla seeds are linolenic ($C_{18:3}$), linoleic ($C_{18:2}$) and oleic ($C_{18:1}$) acids; palmitic ($C_{16:0}$) and stearic ($C_{18:0}$) acids are minor components (Table 3). Perilla oil contains higher unsaturated fatty acids than other vegetable oils, especially high content of linolenic and linoleic acids. Depending on the variety and the growing conditions, the content of the linolenic acid is 54–64%, which

Table 3   Major fatty acid composition of total lipid in Perilla seed (%)

Fatty acids	Shin and Kim (1994)	Longvah and Deosthale (1991)	Tsuyuki et al. (1978)
$C_{16:0}$	6.3 – 6.7	8.92	4.04 – 6.64
$C_{18:0}$	1.5 – 1.7	3.77	0.95 – 1.44
$C_{18:1}$	13.2 – 14.9	12.92	13.12 – 18.34
$C_{18:2}$	14.3 – 17.0	17.61	18.19 – 20.03
$C_{18:3}$	61.1 – 64.0	56.76	53.63 – 58.97
Total saturates	7.9 – 8.4	12.69	4.99 – 8.08
Total Unsaturates	88.6 – 95.9	87.29	84.97 – 97.34
Polyunsaturates	75.4 – 81.0	74.37	71.82 – 79.00

is not less than that of linseed oil and 6 to 8 times more than that found in mustard and soybean oils. The content of linoleic acid in Perilla oil is similar to that of linseed and mustard oils.

Tsuyuki *et al.* (1978) reported that Perilla seed has small amount (0.08–0.27%, individually) of capric acid ($C_{10:0}$), lauric acid ($C_{12:0}$), myristic acid ($C_{14:0}$), eicosanoic acid ($C_{20:0}$), etc.

The study on the fatty acid composition among major lipid classes of Perilla seed total lipids shows that the fatty acid profile of NL is similar to that of the total lipids; fatty acid profile of GL and PL are high in palmitic, stearic and linoleic acids compared to the neutral lipid fraction. The GL and PL fractions of Perilla seed total lipids contain significant amounts of medium-chain fatty acids, i.e., capric and myristic acids (Table 4).

Minor Components

Sterol and tocopherol were reported to be minor components of Perilla oil. Park *et al.* (1982) analysed the sterol composition of oil extracted from seed (*P. frutescens* Britt. var. *crispa* Decaisne). The results showed that the unsaponifiable matter of Perilla oil was composed of more than 60% of 4-desmethyl sterol fraction and trace amounts of 4,4-

Table 4   Fatty acid composition of major lipid classes Perilla seed [a]

Fatty acids	Neutral lipids	Glycolipids	Phospholipids
$C_{10:0}$	–	1.0 – 14.1	3.2 – 15.7
$C_{14:0}$	–	1.4 – 15.2	–
$C_{16:0}$	6.3 – 6.9	0.5 – 13.8	7.7 – 15.9
$C_{18:0}$	1.4 – 1.7	2.7 – 4.4	1.5 – 3.4
$C_{18:1}$	13.0 – 14.9	13.0 – 14.6	6.9 – 14.1
$C_{18:2}$	14.4 – 16.0	14.3 – 18.6	15.3 – 29.3
$C_{18:3}$	62.1 – 64.0	40.2 – 56.1	31.6 – 58.2

[a] From Shin and Kim, 1994.

Table 5 Sterol composition in Perilla oil[a]

Fractions	Campesterol	Stigmasterol	β-Sitosterol	Δ⁵-Avenasterol
Total sterol	4.4 – 6.5	0.3 – 0.8	78.0 – 81.7	12.6 – 14.8
Esterified sterol	tr.	tr.	98.9 – 99.6	tr.
Steryl glycoside	3.5 – 7.5	0.2 – 1.2	79.9 – 84.8	11.6 – 13.5
Free sterol	9.4 – 10.4	3.1 – 10.0	54.7 – 72.6	5.4 – 32.8

[a] From Park *et al.*, 1982.

dimethyl sterol and 4-monomethyl sterol fractions. Table 5 shows the sterol fraction of Perilla oil; esterified sterol, steryl glycoside, free sterol and sterol composition of total sterol fraction. β-Sitosterol occurs in more than 99% of the esterified sterol fraction and 55–93% in the free sterol fraction. Also they mentioned that the sterol composition of the seed oil is varies with its origin.

The total tocopherol content of total lipids extracted from Perilla seeds was 49.1–67.6 mg/100g oil. The γ-form existed as the major tocopherol (92%) with small amounts of the α- and δ-forms and the β-form absent (Shin and Kim, 1994).

In Perilla seed Park *et al.* (1993) identified physiologically active brassinosteroid, 0.50–0.8ng, as brassonolide, per gram of fresh weight, mainly castasterone with some homodolicholide.

The Change of Lipid Composition during Maturation and Germination

Min and Kim (1992a, b) studied the change of lipid composition during maturation of Perilla seed. They found that the content of ether-extractable lipids was increased continuously as the seeds matured while the content of triglyceride, the essential component of ether-extractable lipids, increased rapidly at the beginning of maturation and were 61.4–68.2% in matured seed (30 days after flowering). The contents of glycolipids and phospholipids were reduced and the amount of individual component of glyco- and phospholipids varied irregularly.

Kim *et al.* (1994) compared the lipid composition of germinated and non-germinated Perilla seed. They reported that the amount of triacylglycerol among neutral lipids was reduced, and free fatty acids and diacylglycerol were increased during germination. The contents of phophatidylethanolamine among polar lipids increased significantly and the amount of tocopherol, especially γ-form, increased notably.

THE IMPROVEMENT OF THE OXIDATIVE STABILITY OF PERILLA OIL

The main problem of Perilla oil as edible oil is its high content of linolenic acid, highly unsaturated fatty acid, which is easily oxidized, and thus limiting its use. Thus, the oxidative stability should be improved to minimize the quality deterioration by rancidity.

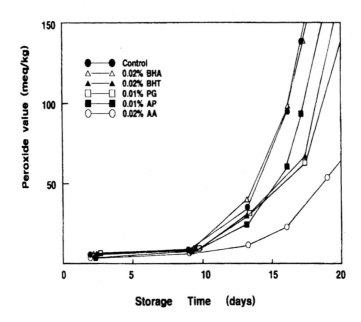

Figure 1   Effect of Butylated hydroxyanisole (BHA), Butylated hydroxytoluene (BHT), Propyl gallate (PG), Ascorbyl palmitate (AP) and Ascorbic acid (AA) on The Oxidation of Perilla oil Stored at 60° C (From Yi and Shin, 1989)

Yi and Shin (1989) reported that rancidity was inhibited more significantly when ascorbic acid in reversed micelle form was added to Perilla oil than when the mixed antioxidants (butylated hydroxyanisole, butylated hydroxytoluene, propyl gallate, ascorbyl palmitate) was added (Figure 1). They also mentioned that δ-tocopherol did not have the synergistic effect on ascorbic acid.

Cha and Choi (1990) found that tocopherol did not improve the oxidative stability since Perilla oil contained 400 ppm of tocopherol originally. In the case of organic acids, the antioxidative effect was increased in the following order; L-ascorbic acid > L-ascorbyl stearate > DL-malic acid > tartaric acid > citric acid.

Ahn *et al.* (1991) reported that under the Rancimat AOM test (97.8 ± 0.2°C) the induction period of Perilla oil was longer than that of purified soybean oil, having 16 hours of AOM time when 5% of commercial lecithin was added to Perilla oil having 2 hours of induction period. As more lecithin was added, the induction period increased. The mixture of tocopherol, citric acid and ascorbyl palmitate showed synergistic effect on lecithin.

Kashima *et al.* (1991) reported the antioxidant effect of phospholipids on the oxidative stability of refined Perilla oil (PO), tocopherol-free Perilla oil (POF) and tocopherol-enriched Perilla oil (POR). The oxidative stability of PO was increased noticeably by adding phosphatidylethanolamine (PE) and phosphatidylserine (PS). However, the addition of PE and PS did not have the antioxidant effect on POF. The oxidative stability of POR was lower than that of PO even though it had higher tocopherol, and was also increased by adding PE and PS. The reason that phospholipids showed antioxidant effect on Perilla oil was due to its synergistic effect with existing tocopherol.

Kim *et al.* (1994) reported that Perilla oil extracted from germinated Perilla seeds was more stable for oxidation than that extracted from nongerminated Perilla seeds. The oxidative stability was significantly increased in the germinated seeds after being stored more than one year whereas a little increased in the fresh seeds.

NUTRITIONAL VALUE

The nutritional value of Perilla oil is somewhat different from that of other plant oils. Perilla oil contains especially higher content of linoleic and linolenic acids than other vegetable oils. Therefore, Perilla oil is a good source of essential fatty acids and can be used to optimize fatty acid ratio of n-6 and n-3 by mixing it with other vegetable oils. Essential fatty acids are those that are required by human body for normal growth, maintenance and for normal physiological functions, and are those that are not endogenously synthesized or those that are synthesized only in insufficient quantities. Therefore they must be supplied in the diet. Linoleic acid, a n-6 fatty acid, and α-linolenic acid, a n-3 fatty acid, are life-saving fatty acids required for phospholipid components of cell membrane, nuclear membrane and mitochondrial membrane. Since the two fatty acids are in a competitive relationship due to their structural similarity, it is harmful unless the two fatty acids are in proper ratios in the diet. Several studies have been published (Bang *et al.*, 1980; Dyerberg, 1986) showing that α-linolenic acid is closely related to anti-hypertensive effect and anti-thrombosis.

Kim and Kim (1989) reported that as dietary Perilla oil was increased, the concentrations of serum cholesterol and triglyceride was decreased. Other workers (Park *et al.*, 1992; Chung *et al.*, 1986; Nam *et al.*, 1981; Park and Han, 1976) reported similar results. Lee *et al.* (1987) reported that Perilla oil had effects on reducing serum cholesterol. Since below 15% dietary intake level of Perilla oil influences the cellular immune response ability, it is useful for cardiovascular disease or immune response. Therefore, for reducing serum cholesterol and improving immune response, it is recommended to diet with unsaturated fatty acid, such as is available from Perilla oil.

Han *et al.* (1983) reported that ingested Perilla oil was transformed to eicosapentaenoic acid (EPA) which restrains formation of thromboxane A_2 (TXA_2) and affects antithrombosis. They reported that when rats were fed *ad libitum* a diet containing 15% of Perilla oil for 15 weeks, the bleeding time was significantly delayed. Also the amount of malondialdehyde was decreased which is an indicator of TXA_2 and increased EPA and decreased arachidonic acid (AA) in the platelet. According to these results, the ingestion of Perilla oil, like fish oil containing high EPA, may result in the delay of bleeding time due to thrombogenesis reduction by increasing the ratio of EPA/AA in the platelet. Therefore, the intake of Perilla oil may be recommended for protecting chronic diseases. But, Suh and Cho (1980) reported that the formation of C_{20}–C_{22} (n-3) acids was not activated by elongation of $C_{18:3}$ (n-3) in Perilla oil in the animal body.

Table 6   Amino acid composition (mg/gN) of *Perilla frutescens* seed protein compared to that of FAO whole egg protein

Amino acid	Longvah and Deosthale (1991)	Standall et al. (1985)	FAO whole egg protein[a]
Threonine	181	182	320
Valine	174	113	428
Cystine	89	83	152
Methionine	174	86	210
Isoleucine	234	100	393
Leucine	374	363	551
Tyrosine	244	145	260
Phenylalanine	311	228	358
Lysine	240	221	436
Aspartate	556	513	601
Serine	443	343	478
Glutamate	1423	1358	796
Proline	308	376	260
Glycine	340	238	207
Alanine	296	297	370
Histidine	203	180	152
Arginine	807	143	381
Tryptophan	-	78	93
Essential amino acids	2021	1185	3201
Total amino acids	6397	5733	6446

[a]From FAO(1978)

On the other hand, when Perilla oil ingestion is too much, it may result in formation of lipid peroxides in the body causing the depletion of antioxidant materials (Choi *et al.*, 1987). Lee *et al.* (1976) reported that a deficiency of vitamin E was shown in rats and chicks fed 15% Perilla oil in the diet for 4 weeks. The major symptoms in rats were hair loss around the neck and serious skin lesion and in chicks significant muscle weakness and discoloration of the skin. Kwak and Choi (1992) reported that animals fed Perilla oil showed lower hepatic microsomal lipid peroxidation than animals fed corn oil, and was similar to animals fed tallow even though Perilla oil has a significantly higher polyunsaturated fatty acids/saturated fatty acids (P/S) ratio than corn oil. They suggested that hepatic microsomal lipid peroxidation was influenced not only by the degree of unsaturation of dietary lipids but also by the positions of double bonds in the fatty acids. They also reported that rats fed Perilla oil showed significantly lower prostaglandin E_2 (PGE_2) and thromboxane B_2 (TXB_2) production than rats fed corn oil. Therefore, the intake of Perilla oil may decrease tumorigenesis caused by lipid peroxides and eicosanoids. The effects of Perilla oil on formation of peroxides in the body is still in conflict, as mentioned above. Lee and Cho (1988) reported that the addition of vitamin C, vitamin E, ethylenediamine tetraacetate (EDTA) to a Perilla oil diet can slightly decrease the formation of peroxides in the serum and the tissues.

Table 7 Food intake, protein efficiency ratio (PER) and net protein ratio (NPR) in groups of rats fed casein and Perilla protein (Mean ± SE of six animals in each group)[a]

Diet group	Protein (%)	Food intake (g/4 weeks)	Body weight gain (g/4 weeks)	PER	NPR
Casein	10	268 ± 8.8	77.7 ± 3.2[b]	2.99 ± 0.88	3.67 ± 0.04
P. frutescens	10	255 ± 11.9	49.0 ± 2.2[b]	2.07 ± 0.09[b]	2.87 ± 0.11[b]
Protein-free	-	103 ± 6.2	18.8 ± 3.1	-	-

[a]From Longvah and Deosthale, 1991.
[b]Significantly different at 0.1% level.

Since the antioxidative effect of Perilla oil is insufficient because of the scanty amount of tocopherol in Perilla oil, supplementation of an appropriate amount of antioxidants, including tocopherol, etc., will enable the reduction in the formation of lipid peroxides. More research should be performed to elucidate the appropriate P/S ratio, the appropriate n-6/n-3 ratio, the formation of lipid peroxides and the prevention of their formation, and the effect on metabolism when Perilla oil is ingested.

In addition to dietary oil, oil seeds generally contain a high content of beneficial protein. Perilla seed contains 18–28% (average 23%) of protein, and the residue left after oil extraction can be used as a protein source for humans and animals. The amino acid composition of Perilla seed protein is shown in Table 6. Amino acid composition varies with the species, the environmental conditions of cultivation, and the method of analysis used. Compared to whole egg, the total essential amino acid content of Perilla protein is very low. Longvah and Deosthale (1991) reported the limiting amino acid of Perilla seed protein was valine; however, Standall et al. (1985) reported it was isoleucine. Table 7 shows the body weight gain, protein efficiency ratio (PER), and net protein ratio (NPR) after animal feeding with Perilla seed protein. According to this result, there was no significant difference in feed intake between animals fed Perilla seed protein and with casein protein. However, body weight gains for animals fed casein protein and Perilla seed protein were 78g/4 weeks and 49g/4 weeks, respectively. PER value of casein protein, 2.99, was significantly higher than that of Perilla seed protein, 2.07. The protein digestibility of these proteins showed similar trends. These results suggested that the quality of Perilla seed protein was poor in terms of absorptiveness and total content of essential amino acids. However, there were no significant quality differences among sesame and other oilseed proteins.

PHYSIOLOGICAL AND THERAPEUTICAL ROLES

Anticarcinogenic Effect

Many research papers have been published on the physiological roles of eicosapentaenoic acid (EPA) and docosahaxaenoic acid (DHA) which exist abundantly in marine oils (Carrol, 1986; Isoda et al., 1988). α-Linolenic acid, which exists abundantly in soybean oil, rapeseed oil and Perilla oil, is the same n-3 fatty acid as DHA and EPA. Until recently,

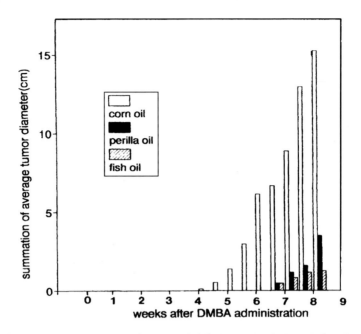

Figure 2   Average Tumor Diameter after DMBA Administration in Rats Fed on Diets Containing Different Fats (From Yonekura and Sato, 1989)

however, the physiological roles of the α-linolenic acid had not been extensively studied, and thus, the essentiality of α-linolenic acid had not been clear. Among the food components, lipid is the most extensively studied component for the relationship between its consumption and cancer risk. Carroll *et al.* (1981, 1984) reported that there was a significant positive relationship between fat intake and mortality from breast, pancreas, colon, rectum, prostate or ovary cancer. The relationship between the type of fatty acids consumed and the risk of cancer has not been fully understood. However, many researchers have reported that high intake of n-6 type fat (linoleic acid) enhanced the carcinogenecity, and that intake of n-3 type fat such as linolenic acid, EPA or DHA lowered breast and colon cancer risk.

Yonekura and Sato (1989) studied the inhibitory effects of dietary Perilla oil, fish oil and corn oil on the breast cancer in rats. The authors reported that Perilla oil and fish oil showed significant inhibitory effect on the dimethylbenzanthracene (DMBA) induced breast cancer in rats (Figure 2).

Recently, several researchers reported that α-linolenic acid had inhibitory effects on breast cancer. Cameron *et al.* (1989) studied the effects of dietary fats on the inhibition of DMBA induced cancer risk in rats by feeding the rats with diets containing 6% either linseed oil (α-linolenic acid, about 50%), fish oil (EPA, 13%), lard, evening-primrose oil (γ-linolenic acid, 9%), or safflower oil (linoleic acid, 78%). They reported that linseed oil and fish oil significantly suppressed the DMBA induced cancers in rats. It was interesting that inhibitory activity of linseed oil which contained 50% of α-linolenic acid was

Table 8 Methyl nitrosourea-induced large-bowel tumors in CD-Fischer rats fed with various fat diets[a]

Dietary[b] group	Effective no. of rats	No. of rats with tumors	No. of tumors per rat
sf	26	12 (46%)	0.6 ± 0.1[c]
SF	25	14 (56%)	0.8 ± 0.2
PR	26	5 (19%)[d]	0.2 ± 0.1[d]
PL	26	15 (58%)	0.9 ± 0.2

[a]From Narisawa et al., 1990.
[b]Rats receiving an intrarectal dose of 2mg of methyl nitrosourea 3 times a week for 2 weeks were fed with diet containing 5%(sf group) or 12%(SF group) safflower oil, 12% perilla oil (PR group), or 12% palm oil (PL group). The experiment was terminated at week 36.
[c]Mean ± SEM
[d]Significantly different from other groups, $p < 0.05$ or 0.01 by X^2 test and Student's t-test.

significantly higher than that of fish oil which contained EPA and DHA. It was also noted that the cancer development in the groups treated with linseed oil or fish oil along with cancer inducing agent DMBA was significantly lower than that in the control which was treated with lard but no DMBA.

Narisawa et al. (1990) examined the methyl nitrosourea (MNU, cancer inducing chemical) treated rats after they had been fed for 35 weeks with diets containing 12% of either Perilla oil, palm oil or safflower oil (Table 8). Perilla oil treatment clearly suppressed large intestine cancer. The number of rats with cancer and the number of cancers per rat in the Perilla oil treated group were about one third and one fourth of other groups, respectively.

Park et al. (1993) reported that Perilla oil and fish oil had similar inhibitory effects on the chemically (N-methyl-N'-nitro-N-nitrosoguanidine, MNNG) induced colon cancer in rats. The author explained that Perilla oil and fish oil inhibited the colon cancer because the oil intake affected the content of arachidonic acid, a precursor of TXA_2 and PGE_2. There are some other published papers on the inhibitory effects of α-linolenic acid on carcinogenecity (Hori et al., 1987; Fritsche, 1988); these are not detailed here.

It is worthwhile to note that α-linolenic acid has inhibitory effects on mammary and large intestine cancer risk. It is not an exaggeration to say that in these days humans are constantly exposed to many different kinds of carcinogens derived from the polluted environments. It is thus expected that daily intake of Perilla seed or Perilla oil may reduce the cancer risk. How to use the seed or oil as food ingredient might be an important matter for healthy diet in the future. Cancer death in Japan is reportedly higher than in Korea. Although the reasons for higher cancer death in Japan are not simple, it can be assumed that it might be, at least to some extent, due to the westernized Japanese diets with high intake of linoleic acid. Koreans still relatively well observe traditional diet habits and eat Perilla seed and oil.

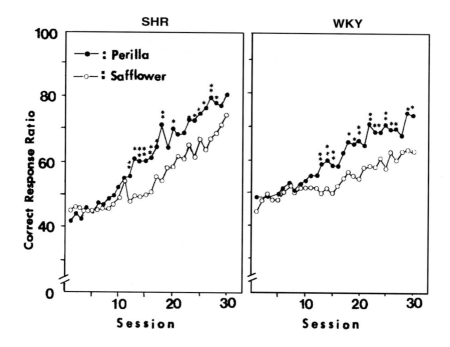

Figure 3   Correct Response Ratio in The Brightness-discrimination Learning Test. (From Yamamoto *et al.*, 1987)

Effects of Brain and Nerve System

Attention has been drawn to the effects of α-linolenic acid on the learning ability since it was reported that this acid was an essential fatty acid in the nerve system. Yamamoto *et al.* (1987) fed SHR rats and WKY rats for 2 generations with linoleic acid rich feed (safflower oil, 5%) or α-linolenic acid rich feed (Perilla oil, 5%), then carried out a brightness-discrimination learning test. The authors reported that dietary Perilla oil group in SHR rat and WKY rat showed better correction rate after 10 time tests (1 per day) and it showed the significant difference after 15 time tests (Figure 3). The increased learning ability by the intake of α-linolenic acid might be due to high concentration of induced DHA in brain from α-linolenic acid. Bourre *et al.* (1989) also reported that α-linolenic acid increased learning ability. They fed female Wister rat for 2 generations with feeds containing sunflower oil (no linolenic acid) or soybean oil (1% linolenic acid). Rats from third generation were used for the experiments. The result showed that sunflower oil diet significantly lowered the DHA level in the brain and soybean oil diet significantly increased the DHA level in the brain. The results also showed the positive relationship between the level of DHA in the brain and the α-linolenic acid contents in diet during brain developing periods. The authors also reported that dietary α-linolenic acid affected the visual nerve system as a result of a learning ability test in a shuttle box. Low intake of dietary α-linolenic acid led to abnormal symptoms in electroretinogram. Neuringer *et al.* (1984, 1986) fed prenatal and postnatal infant rhesus monkey with α-linolenic acid deficient diets. They found that DHA contents in the retina and in the brain of the α-linolenic acid deficient group were 1/2 and 1/4 of those in the control group, respectively. Thus, the sight of infant monkey with α-linolenic acid deficiency was failing.

Effects of Survival Time

Shimokawa *et al.* (1988) reported that α-linolenic acid rich diets increased life span of rats. They fed hypertensive rats with diets containing 5% of Perilla oil (65% of α-linolenic acid) or 5% of safflower oil (no linolenic acid) and observed the life span of the rats. The results showed that the mean life spans of the dietary Perilla oil group and the dietary safflower oil group were 59.5 weeks and 50.9 weeks, respectively. That is, the life span of the dietary Perilla oil group was about 17% longer than that of the dietary safflower group.

Renaud *et al.* (1983, 1981) let 25 French farmers, who had been given a saturated fat rich diet, substitute butter with margarine made of canola oil (10% of α-linolenic acid) for 1 year and determined the serum lipids and platelet aggregation. By substituting the butter with margarine, the percent of α-linolenic acid in the farmers' diets increased from 1.2% to 3.5%. One year after, the determined EPA content in platelet and serum of the farmers was slightly increased, but the platelet aggregation had decreased significantly. Renaud *et al.* (1983) reported that even though the regular intake of α-linolenic acid did not greatly increase the serum EPA content, it was important in the prevention of cardiovascular disease and thrombosis. They called it "small is beautiful" and recommended regular intake of α-linolenic acid.

REFERENCES

Ahn, T.H., Kim, J.S., Park, S.J. and Kim, H.W. (1991) Antioxidative effect of commercial lecithin on the oxidative stability of Perilla oil. *Korean J. Food Sci. Technol.* **23**, 251–255 (in Korean).

Bang, B.O., Dyerberg, J. and Sinclair, H.M. (1980) The composition of the Eskimo food in north western Greenland. *Am. J. Clin. Nutr.*, **33**, 2657–2661.

Bjerve, K.S., Fischer, S., Wammer, F. and Egeland, T. (1989) α-Linolenic acid and long-chain n-3 fatty acid supplementation in three patients with n-3 fatty acid deficiency : effect on lymphocyte function, plasma and red cell lipids, and prostanoid formation. *Am. J. Clin. Nutr.,* **49**, 290–300.

Bourre, J.M., Francois, M., Youyon, A., Dumont, O., Piciotti, M., Pascal, G. and Durand, G. (1989) The effects of dietary α-linolenic acid on the composition of nerve membranes, enzymatic activity, amplitude of electrophysiological parameters, resistance to poisons and performance of learning tasks in rats. *J. Nutr.*, **119**, 1886–1892.

Cameron, E., Bland, J. and Marcuson, R. (1989) Divergent effects of n-6 and n-3 fatty acids on mammary tumor development in C3H/Heston mice treated with DMBA. *Nutr. Res.*, **9**, 383–393.

Carrol, K.K. (1984) Role of lipids in tumorigenesis. *J. Am. Oil Chem. Soc.*, **61**, 1888–1891.

Carrol, K.K. (1986) Biological effects of fish oils in relation to chronic disease. *Lipids*, **21**, 731–732.

Carrol, K.K., Hopkins, G.J., Kennedy, J.G. and Davidson, M.B. (1981) Essential fatty acids in relation to mammary carcinogenesis. *Prog. Lipid Res.*, **20**, 685-690.

Cha, G.S. and Choi, C.U. (1990) Determination of oxidative stability of Perilla oil by the Rancimat method. *Korean J. Food Sci. Technol.*, **22**, 61–65 (in Korean).

Chang, G.H. (1995) *History of the use of dietary lipids in Korea. Soo Hak Sa*, Seoul, pp.143–215 (in Korean).

Choi, K.W., Park, M.H. Chang, K.S. and Cho, S.H. (1987) Effect of dietary fish oil on lipid peroxidative and antiperoxidative system in rat liver and brain. *J. Korean Soc. Food Nutr.*, **16**, 147–155 (in Korean).

Chung, S.Y., Seo, M.H., Park, P.S., Kanji, J.S. and Kanji, J.O. (1986) Influences of dietary fats and oils on concentration of lipids in serum and liver of rats on hypercholesterolemic diet. *J. Korean Soc. Food Nutr.*, **15**, 75–81 (in Korean).

Dyerberg, J. (1986) Linolenate-derived polyunsaturated fatty acids and prevention of atherosclerosis. *Nutr. Rev.*, **44**, 125–134.

FAO Nutritional Studies No. 24. (1978) Amino Acid Composition and Biological Data on Proteins. Food and Agricultural Organization of the United Nations, Rome.

Fritsche, K.L. (1988) Reduced growth and metastasis of a transplantable syngenic mammary tumor by dietary α-linolenic acid. *J. Am. Oil Chem. Soc.*, **65**, 509.

Han, Y.N., Yoon, H.W., Kim, S.H. and Han, B.M. (1987) Effects of Perilla oil intake on bleeding time, thromboxane formation and platelet fatty acid in rats. *Korean J. Pharmacogn.* **18**, 513 (in Korean).

Holman, R.T., Johnson, S.B. and Hatch, T.F. (1982) A case of human linolenic acid deficiency involving neurological abnormalities. *Am. J. Clin. Nutr.*, **35**, 617–623.

Hori, T., Moriuchi, A., Okuyama, H., Sohajima, T., Koizumi K., and Kojima, K. (1987) Effect of dietary essential fatty acids on pulmonary metastasis of ascites tumor cells in rat. *Chem. Pharm. Bull.*, **35**, 3925–3927.

Isoda, Y. and Hirano, J. (1988) Cancer and lipids. *Hygienic Chem.*, **34**, 295–302 (in Japanese).

Kashima, M., Cha, G.S., Isoda, Y., Hirano, J. and Miyazawa, T. (1991) The antioxidant effects of phospholipids on Perilla oil. *J. Am. Oil Chem. Soc.*, **68**, 119–122.

Kim, C.K., Song, G.S., Kwon, Y.J., Kim, I.S. and Lee, T.K. (1994) The effect of germination of Perilla seed on the oxidative stability of the oil. *Korean J. Food Sci. Technol.*, **26**, 178–183 (in Korean).

Kim, W.K. and Kim, S.H. (1989) The effect of sesame oil, Perilla oil and beef tallow on body lipid metabolism and immune response. *Korean J. Nutr.*, **22**, 42–53 (in Korean).

Kwak, C.S. and Choi, H.M. (1992) Effects of intake of Perilla oil or corn oil and 2-acetyl-aminofluorene treatment on lipid peroxidation, PGE2 and TXB2 production in rats. *Korean J. Nutr.*, **25**, 351–359 (in Korean).

Lee, I.S. and Cho, C.S. (1988) Effect of antioxidants added Perilla oil diet on serum and tissue in rats. *Korean Oil Chem. Soc.*, **5**, 29–38 (in Korean).

Lee, J.M., Kim, W.Y. and Kim, S.H. (1987) A study of Korean dietary lipid sources on lipid metabolism and immune function in rat. *Korean J. Nutr.*, **20**, 350–366 (in Korean).

Lee, Y.C., Kwak, T.K. and Lee, K.Y. (1976) Relationship between vitamin E and polyunsaturated fat. A comparative animal study emphasizing Perilla seed oil as a fat constituent. *Korean J. Nutr.*, **9**, 283–291 (in Korean).

Longvah, T. and Deosthale, Y.G. (1991) Chemical and nutritional studies on Hanshi (*Perilla frutescens*), a traditional oilseed from northeast India. *J. Am. Oil Chem. Soc.*, **68**, 781-784.

Min, Y.K. and Kim, Z.U. (1992a) Change of glycolipids and phospholipids during maturation of Perilla seed (*Perilla frutescens*). *J. Korean Agric. Chem. Soc.*, **35**, 146–151 (in Korean).

Min, Y.K. and Kim, Z.U. (1992b) Change of lipids during maturation of Perilla seed (*Perilla frutescens*). *J. Korean Agric. Chem. Soc.*, **35**, 139–145 (in Korean).

Nam, H.K., Sung, H.C. and Chang, I.Y. (1981) Studies on the effect in degree of saturation of fats on serum cholesterol level in the rabbit. *J. Korean Soc. Food Nutr.*, **10**, 27–37 (in Korean).

Narisawa, T., Takahashi., M., Kusaka, N., Yamazaki, Y., Koyama, H., Kotanaga, H., Nishizawa, Y. and Kotsugai, M. (1990) Inhibition of large-bowel carcinogenesis in rats by dietary Perilla oil rich in the n-3 polyunsaturated fatty acid α-linolenic acid. *Ishiyaku*, **153**, 103–104 (in Japanese).

Neuringer, M., Connor, W.E., Lim, D.S., Barstad, L. and Luck, S. (1986) Biochemical and functional effects of prenatal and postnatal n-3 fatty acid deficiency on retina and brain in rhesus monkeys. *Proc. Natl. Acad. Sci.*, **83**, 4021–4025.

Neuringer, M., Connor, W. E., Petten, C.V. and Barstad, L. (1984) Dietary n-3 fatty acid deficiency and visual loss infant Rhesus monkey. *J. Clin. Invest.*, **73**, 272–276.

Park, H.S. and Lee, S.M. (1992) Effects of dietary n-3 fatty acid and fat unsaturation on plasma lipids and lipoproteins in rats. *Korean J. Nutr.*, **25**, 555–568 (in Korean).

Park, H.S., Kim, J.G. and Cho, M.J. (1982) Chemical compositions of *Perilla frutescens* Britton var. *crispa* Decaisne cultivated in different area of Korea. *J. Korean Agric. Chem. Soc.*, **25**, 14–20 (in Korean).

Park, H.S., Kim, J.G. and Hyun K.H. (1983) Brassinosteroid substances in immature *Perilla frutescens* seed. *J. Korean Agric. Chem. Soc.*, **36**, 197–201 (in Korean).

Park, H.S., Seo, E.S., Song, J.H. and Choi, C.U. (1993) Effects of Perilla oil rich in α-linolenic acid on colon tumor incidence, plasma thromboxane B2 level and fatty acid profile of colonic mucosal lipids in chemical carcinogen-treated rats. *Korean J. Nutr.*, **26**, 829–838 (in Korean).

Park, K.R. and Han, I.K. (1976) Effects of dietary fats and oils on the growth and serum cholesterol content of rats and chicks. *Korean J. Nutr.*, **9**, 59–67 (in Korean).

Park, Y.H., Kim, D.S. and Chun, S.J.(1983) Triglyceride composition of Perilla oil. *Korean J. Food Sci. Technol.*, **15**, 164–169 (in Korean).

Renaud, S. and Nordoy, A (1983) " Small is beautiful " α-linolenic acid and eicosapentaenoic acid in man. *The Lancet*, **21**, 1169.

Renaud, S., Moragain, R., Godsey, F., Domont, E., Symington, I.S., Gillanders, E.M. and Obrine, J. (1981) Platelet functions in relation to diet and serum lipids in British farmers. *Br. Heart J.*, **46**, 562–570.

Shimokawa, T. and Okuyama, H. (1988) Effect of dietary α-linolenate/linoleate balance on mean survival time, incidence of stroke and blood pressure of spontaneously hypertensive rats. *Life Science*, **43**, 2067–2075.

Shin, H.S. and Kim, S.W. (1994) Lipid composition of Perilla seed. *J. Am. Oil Chem. Soc.*, **71**, 619–622.

Sonntag, N.O.V. (1979) Composition and characteristics of individual fats and oils. In Swern D, (ed.), *Bailley's Industrial Oil and Fat Products*, John Wiley & Sons, New York, pp. 434–435.

Standall, B.R., Ako, H. and Standall, G.S.S. (1985) Nutrient content of tribal foods from India : *Flemingia vestita* and *Perilla frustescens*. *J. Plant Foods*, **61**, 1471–1453.

Suh, M. and Cho, S.M. (1986) Effect of dietary n-3 fatty acids on mitochondrial respiration and on lipid composition in rat heart.*Korean Biochem. J.*, **19**, 160–167 (in Korean).

Tsuyuki, H., Itoh, S. and Nakatsukasa, Y. (1978) Studies on the lipids in Perilla seed. Research Division in Agriculture, Nihon University, **35**, 224–230 (in Japanese).

Vaugham, J.G. (1970) *The Structure and Utilization of Oil Seeds*.Chapman and Hall LTD., London, p. 120.

Yamamoto, N., Saitoh, M., Moriuchi, A., Nomura, M. and Okuyama, H. (1987) Effect of dietary α-linolenate/linoleate balance on brain lipid compositions and learning ability of rat. *J. Lipid Res.*, **28**, 144–151.

Yi, O.S. and Shin, H.K. (1989) Antioxidative effect of ascorbic acid solubilized via reversed micelle in Perilla oil. *Korean J. Food Sci.Technol.*, **21**, 706–709.

Yonekura, I. and Sato, A. (1989) Inhibitory effects of Perilla and fish oil on 7,12-dimethylbenz[a]anthracene induced mammary tumorigenesis in Sprague-Dawley rats. *Ishiyaku*, **150**, 233–234 (in Japanese).

10. CHEMICAL STUDIES ON THE CONSTITUENTS OF *PERILLA FRUTESCENS*

TOMOYUKI FUJITA and MITSURU NAKAYAMA

Applied Biological Chemistry, College of Agriculture, Osaka Prefecture University, 1-1 Gakuen-cho, Sakai, Osaka 593, Japan

INTRODUCTION

The chemical constituents of Perilla concerning volatile components and pigments have been investigated in detail, as also other compounds having some biological or pharmacological activities, since the plant has been used for various purposes. Its constituents, terpenoids, phenolics, flavonoids, cyanogenic glycosides, and anthocyanins have been reported. In recent years, we have focused on the glycosidic constituents of Perilla, and have isolated about twenty glycosides including nine new glucosides. Additionally, the inhibitory effect of perillosides A and C, and of related monoterpene glucosides on aldose reductase and their structure-activity relationships have also been elucidated.

In this review, we summarize the chemical constituents of Perilla and their biological activities, but the survey is restricted to secondary metabolites. The constituents classified as pigments such as flavonoids and anthocyanins are not discussed here, but will be covered in a later chapter.

CHEMICAL STRUCTURES OF CONSTITUENTS

Since the various beneficial properties ascribed to Perilla are associated with consumption of the leaves and seeds of the plant, their chemical constituents have been thoroughly studied. The constituents have been obtained by steam distillation and extraction with some solvent. The characterization of the volatile components was generally achieved by GC, GC-FT/IR, and GC-MS measurements without purification. The structure elucidation of the compounds, on the other hand, was performed by spectral and chemical evidence, after purification by fine distillation and swelling chromatographic methods such as CC, GC, TLC, and HPLC. Spectroscopical analysis (IR, UV, MS, and NMR, etc.) was used to elucidate the chemical structure of the purified compound. High resolution MS, MS/MS and NMR spectroscopic techniques provided significant information regarding the structural linkages. Accordingly, we present that the constituents of Perilla are separated into volatile and non-volatile components, the latter being further divided into chemical classes, e.g., terpenoids, phenolics, flavonoids, glycosides, and other constituents.

Table 1   Components of the essential oils from the Perilla species

Compounds	Species*	Plant organ	References
(-)-perillaldehyde (1), (-)-limonene (2) α-pinene, etc.	shiso and ao-jiso		Okuda (1967)
elsholtziaketone (13), naginataketone (14), perillaketone (15) (isoamyl-3-furylketone)	egoma		*Ibid.*
citral (9), perillene	lemon-egoma		*Ibid.*
1 (about 50 %), 2, perillyl alcohol (3), pinene, camphene, etc.	ao-jiso (commercial oil)	aerial part	Masada (1975)
13, 14, linalool (6), 1-octen-3-ol, etc.	egoma	aerial part	Fujita *et al.* (1966)
β-caryophyllene (7), elemicin (10), myristicin (11), dillapiole (12), isoegomaketone (16), etc.	several species	aerial part	Ito (1966, 1968)
α-farnesene (8), allofarnesene	ao-jiso	aerial part	Sakai and Hirose (1969)
1 - 3, 6, *trans*-shisool (4), *cis*-shisool (5), etc.	shiso and ao-jiso	aerial part	Fujita *et al.* (1970a, b)
15, 16, etc.	shiso	leaf	Ina and Ogura (1970), Ina and Suzuki (1971)
1, 3, 7, 10, carvone, phenethyl alcohol, etc.	shiso, katamen-jiso	fruit	Kameoka and Nishikawa (1976)
perillaketone (15)	Tennessee	aerial part	Wilson *et al.* (1977)
rosefuran, 7, 15, etc.	Bangladesh	aerial part	Misra and Husain (1987)
Chemotaxonomy 1 – 7, 9 – 16, etc.	110 samples in Japan	aerial part	Ito (1970)
1, 2, 9 – 17, perillene, etc.	a number of species (genetic studies)	leaf, fruit, cotyledon, calyx	Koezuka *et al.* (1984, 1986a, b, c), Nishizawa *et al.* (1990a, b), Yuba *et al.* (1992), Honda (1994)

*shiso, ao-jiso, egoma, lemon-egoma, and katamen-jiso: called in Japanese.

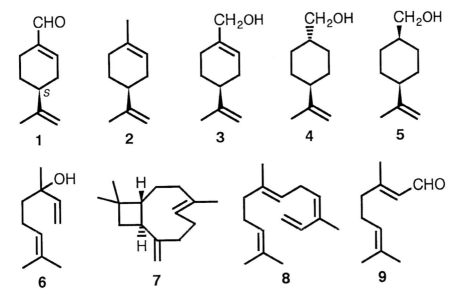

Figure 1   Mono- and sesquiterpenoids in Perilla essential oils

Volatile Components in Essential Oils

The volatile components in Perilla leaves have been obtained as an essential oil by steam distillation. There are many reports on the essential oils obtained from a number of *Perilla* varieties (Table 1). Their volatile compounds, mono- and sesquiterpenes, phenylpropanoids, and furylketones have already been reported, but all of them are not contained in any one subspecies or variety of Perilla. Some typical ones are shown in Figures 1 and 2. Attempts to classify them into the chemotypes of the essential oils have been undergone, as the volatile components in some varieties of Perilla might be characterized by their biosynthesis. It has recently become apparent that the chemotypes of the essential oils can be classified into six types on an individual level, based on the main volatile components (Honda, 1994). These investigations on the chemotypes of the essential oils and on the genetic controls of the volatile compounds will be described in the next chapter. In this section, the chemical structures of the compounds in essential oils are described, even if most of them reported in the past might be obtained from a mixture of different chemotype species.

Commercial Perilla oil (Perilla essential oil), which has a Perilla-like odor, is mainly obtained from green Perilla leaves, called "ao-jiso" in Japanese, by steam distillation, although there are several kinds of Perilla species being cultivated in Japan. Its volatile components are comprised of a rich mixture of mono- and sesquiterpenes. Typical monoterpenes are (-)-perillaldehyde (= (4*S*)-1,8-*p*-menthadien-7-al, (**1**) and (-)-limonene (= (4*S*)-1,8-*p*-menthadiene, (**2**), sesquiterpenes β-caryophyllene (**7**) and α-farnesene (**8**) (Masada, 1975). The major compound, perillaldehyde about 50 – 60 % of the essential

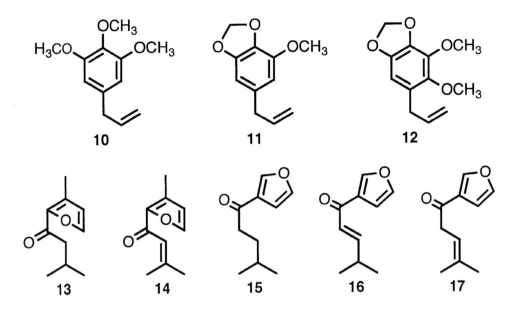

Figure 2   Phenylpropanoids and furylketones in Perilla essential oils

oil, has a powerful fatty-spicy, oily-herbaceous odor and a sweet-herbaceous taste (Arctander, 1969), and it is well-known that its anti-oxime is about two thousand times sweeter than sucrose. Other characteristic compounds having a Perilla-like odor, (-)-perillyl alcohol (**3**), *trans*-shisool (**4**), *cis*-shisool (**5**), and linalool (**6**) are present in Perilla leaves, which also contain α-pinene, β-pinene, camphene, 3-octanol, 1-octen-3-ol, allofarnesene, β-farnesene, etc., as minor components (Table 1). This type is classified as a perillaldehyde type.

Other types of essential oils, in contrast, have little or no Perilla-like odor. They have been classified into five types by Honda (1994). In a type containing phenylpropanoids, elemicin (**10**), myristicin (**11**), and dillapiole (**12**), are present as major compounds. In two other types containing furylketones, either elsholtziaketone (**13**) and naginataketone (**14**), or perillaketone (**15**), isoegomaketone (**16**), and egomaketone (**17**), are also present in the leaves of the plant. The last type being classified to perillene type, contains perillene together with **9** as a minor component (Nishizawa *et al.*, 1989). The other type having a citrus odor, which is classified as *P. frutescens* Britt. var. *citriodora* (Makino) Ohwi, contains citral (**9**) as a major component. These compounds in the Perilla leaves essential oils are also present in the fruits of the same plant (Kameoka and Nishikawa, 1976).

In addition, biosyntheses of mono- and sesquiterpenes have been investigated by using *Perilla* callus, but it is not discussed here (Suga *et al.*, 1986; Tamura *et al.*, 1989; Nabeta *et al.*, 1983, 1984, 1993).

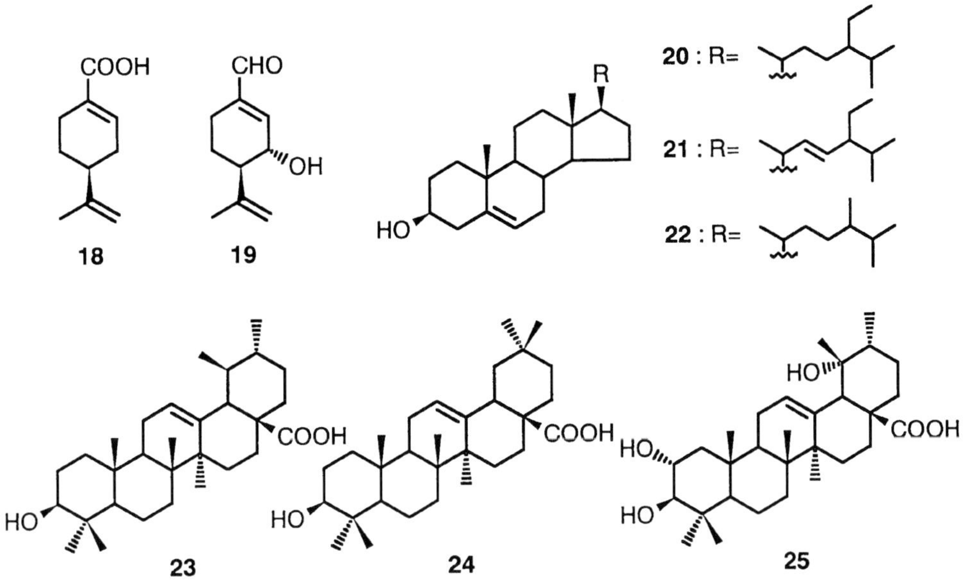

Figure 3 Non-volatile monoterpenes, sterols and triterpenes isolated from *P. frutescens*.

Non-volatile Compounds

The chemical structures of non-volatile compounds are described here. A number of triterpenoids, phenolics, flavonoids, and glycosides, which have various biological activities, have so far been reported. They might be commonly found in some varieties of Perilla in contrast with the volatile components in essential oil.

Terpenoids and Sterols

The structures of non-volatile terpenoids and sterols are shown in Figure 3. As a non-volatile monoterpene, perillic acid (**18**) is well-known as a autooxidation product of the compound **1** (Okuda, 1967). (3*S*,4*R*)-3-Hydroxy-4-(1-methylethenyl)-1-cyclohexene-1-carboxaldehyde (**19**) has been isolated from green Perilla by Matsumoto *et al.* (1995).

Besides the widespread sterols, β-sitosterol (**20**), stigmasterol (**21**), and campesterol (**22**) from both the leaves and seeds of Perilla (Honda *et al.*, 1986; Part *et al.*, 1982; Noda *et al.*, 1975), Perilla leaves also contain triterpenoids, ursolic acid (**23**), oleanolic acid (**24**) (Koshimizu, 1991), and tormentic acid (**25**). The latter acid **25** has been newly identified in purple Perilla leaves, called "aka-jiso" or "shiso" in Japanese, together with the acids **23** and **24** including dillapiole as a major volatile compound in our investigation (Fujita *et al.*, 1994). Studies on biosynthesis of triterpenoids have been carried out by using Perilla cell cultures (Tomita *et al.*, 1985, 1994). Higher terpenoids, five carotenoids, β-carotene, lutein, neoxanthin, antheraxanthin, and violaxanthin, have also been identified in green Perilla leaves by Takagi (1985).

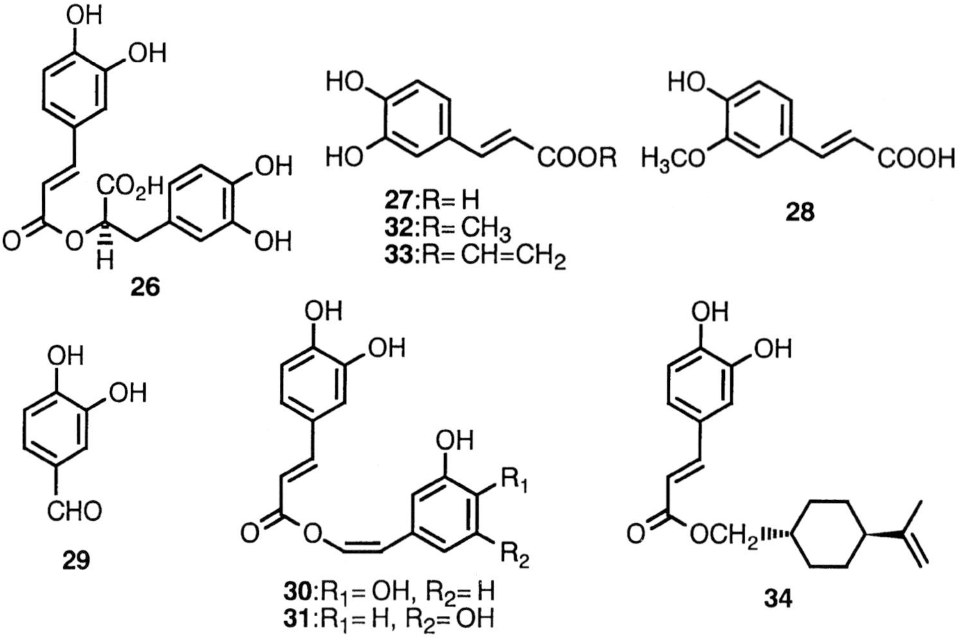

Figure 4   Phenolics and cinnamates from *P. frutescens*.

Phenolics, Cinnamates and Phenylpropanoids

In common with other members of the Labiatae family, Perilla leaves contain a rich mixture of phenolics and cinnamates (Figure 4). Typical of these are cinnamic acid derivatives, rosmarinic acid (**26**), caffeic acid (**27**), and ferulic acid (**28**) (Okuda *et al.*, 1986; Aritomi, 1982). Five esters of caffeic acid, 2-(3,4-dihydroxyphenyl)ethenyl caffeate (**30**), 2-(3,5-dihydroxyphenyl)-ethenyl caffeate (**31**) (Nakanishi *et al.*, 1990), methyl caffeate (**32**), vinyl caffeate (**33**), and 8-*p*-menthen-7-yl caffeate (=*trans*-shisool-3-(3,4-dihydroxyphenyl)-2-propenoate, **34**) (Matsumoto *et al.*, 1995), have also been isolated from *Perilla* leaves. As a phenolic, protocatechuic aldehyde (**29**) is present in Perilla leaves (Fujita *et al.*, 1994). In addition, α- and γ-tocopherol in the leaves of the plant have been identified as an antioxidative substance (Su *et al.*, 1986). Except for the compounds, **10–12** introduced in section 1, small amounts of a phenylpropanoid, eugenol and other aromatics, benzaldehyde, benzylalcohol, and phenethyl alcohol, which have been mainly characterized by the essential oils described in Table 1, are also present.

Flavonoids and Anthocyanins

The characteristic compounds obtained from purple Perilla are introduced here (Figure 5), although there have been many reports of pigments such as flavonoids and anthocyanins from Perilla. Typical flavonoids are apigenin (**35**), luteolin (**36**), scutellarein (**37**), and their glycosides (Ishikura, 1981; Aritomi, 1982; Yoshida *et al.*, 1993), while

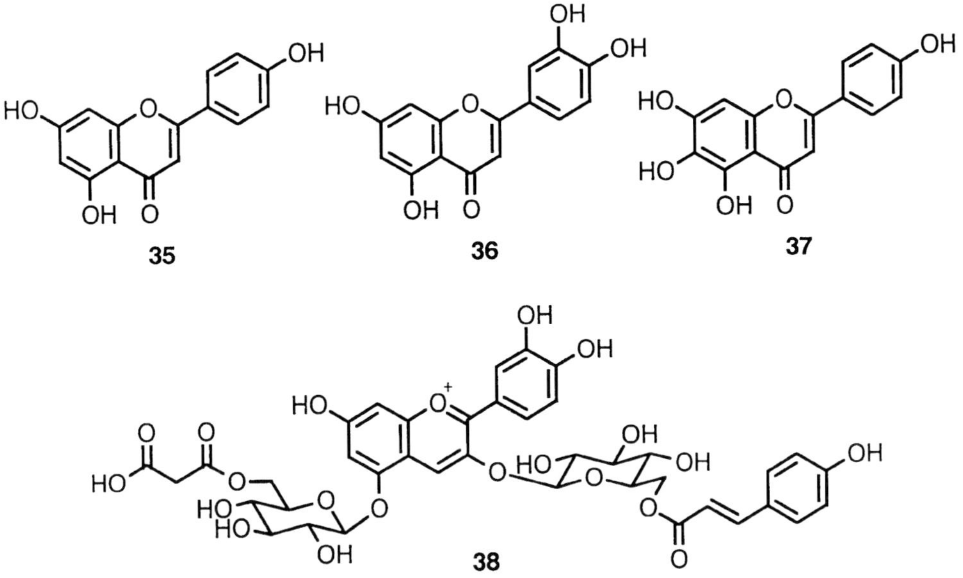

Figure 5 Typical flavonoids and anthocyanin from *P. frutescens*.

typical anthocyanins are acylated glucosides of cyanidin, malonylshisonin (**38**) and shisonin (Kondo *et al.*, 1989 ; Yoshida *et al.*, 1990). Other pigments will be described in the following chapter.

Genetic study on the anthocyanin production in Perilla leaf and stem has been reported by Koezuka *et al.* (1985a), and production of phenylpropanoids and anthocyanins by callus tissue has also been investigated by Tamura *et al.* (1989).

Glycosides

About twenty glycosides, except for pigments, including nine new glucosides, have been found from green and purple Perilla leaves in our investigation, and they are classified into terpenoids, phenylpropanoids, cyanogenics, jasmonoids, phenylvaleric acid and other glycosides, respectively (Figures 6–8).

First, four monoterpene glucosides perillosides A–D (**39–42**), along with eugenyl β-**D**-glucopyranoside (**43**) and benzyl β-**D**-glucopyranoside (**44**), have been isolated from the methanolic extract of green Perilla leaves including perillaldehyde as a major component in the essential oil (Fujita and Nakayama, 1992, 1993). The structure of perilloside A was characterized as (4*S*)- (-)-perillyl β-**D**-glucopyranoside (**39**) by means of spectral and chemical methods. In the same manner, perillosides B, C, and D were determined to be β-**D**-glucopyranosyl (4*S*)-(-)-perillate (**40**), *trans*-shisool-β-**D**-glucopyranoside (**41**), *cis*-shisool-β-**D**-glucopyranoside (**42**), respectively. Two known

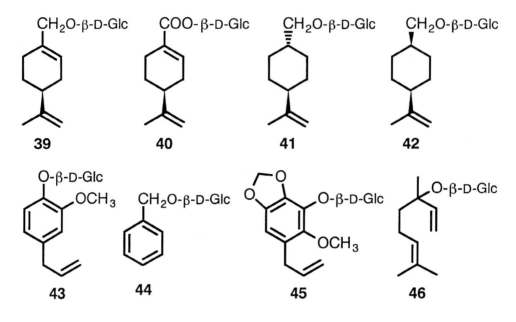

Figure 6   Monoterpenoids and phenylpropanoids glucosides from *P. frutescens*

glucosides, **43** and **44**, have already been isolated from *Melissa officinalis* by Mulkens and Kapetanidis (1988) and from *Carica papaya* fruit by Schwab and Schreier (1988), respectively. The corresponding aglycones of these glucosides **39**-**44** were mentioned above as volatile components. β-Sitosteryl β-**D**-glucopyranoside was also obtained from the same extract of the green Perilla as another terpenoid glucoside.

Next, phenylpropanoid glucoside perilloside E (**45**) along with **43**, **44**, and linalyl β-**D**-glucopyranoside (**46**) are present in the methanolic extract of purple Perilla leaves including dillapiole as a major compound in the essential oil (Fujita *et al.*, 1994). The structure of perilloside E, which has a penta-substituted benzene ring consisting of allyl, methylenedioxy, methoxyl and β-**D**-glucopyranosyloxy moieties, was determined to be 6-methoxy-2,3-methylenedioxy-5-allylphenyl β-**D**-glucopyranoside (**45**) by means of long-range ^{13}C-^{1}H correlations of the NMR measurements. The corresponding aglycone of **45**, 6-methoxy-2,3-methylenedioxy-5-allylphenol was also found for the first time in plants. Its methylation product is identical to dillapiole. These glucosides **39–46** seem to be closely associated with the volatile constituents in the essential oils, as each aglycone could afford related volatile compound. The result on the analysis of the glucosidic composition in the plant should support the genetic study on the volatile component formation of *Perilla* species.

Two cyanogenic glycosides, prunasin (= (*R*)-2-(β-**D**-glucopyranosyloxy) phenylacetonitrile, **47**) and amygdalin isomer, (*R*)-2-(2-*O*-β-**D**-glucopyranosyl-β-**D**-glucopyranosyloxy) phenylacetonitrile, **48**) have been reported by Aritomi *et al.* (1985, 1988). They were hydrolyzed with concentrated hydrochloric acid to (*R*)-mandelic acids.

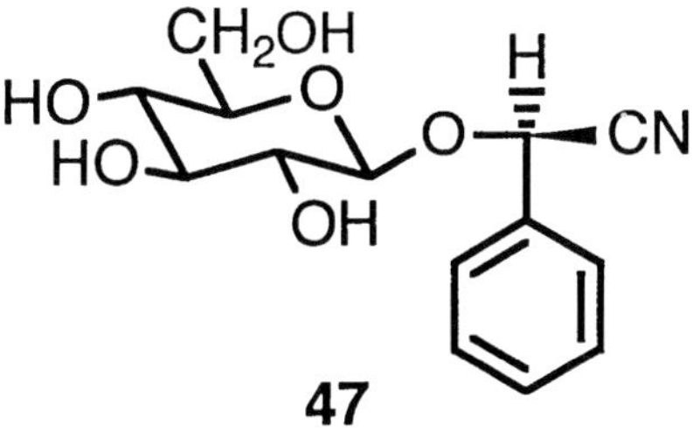

Figure 7 Cyanogenic Glycosides from *P. frutescens*.

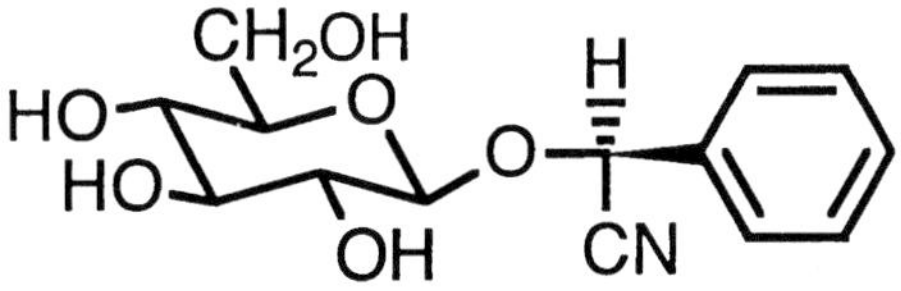

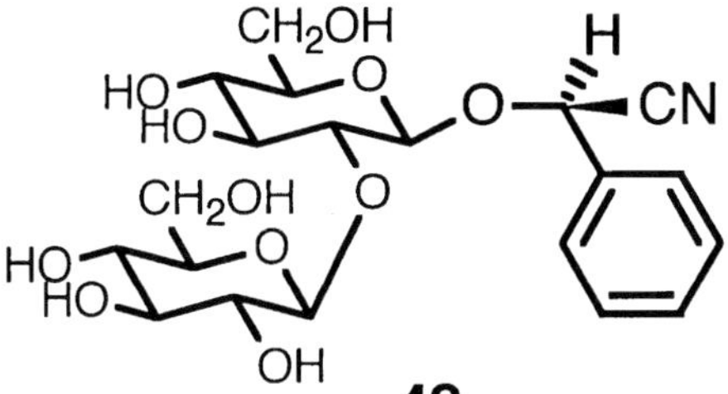

Additionally, sambunigrin (= (*S*)-2-(β-**D**-glucopyranosyloxy) phenyl-acetonitrile, **49**), along with **47** and **48**, is also present in both green and purple Perilla leaves (Fujita *et al.*, 1994). The chemical and physical properties of **47** and **49** are very similar to each other, since these are diastereoisomers of mandelonitrile glucopyranoside (Schwarzmaier, 1976). However, their ^{1}H- and ^{13}C-NMR spectra are distinguishable by comparison of signals at the anomeric and nitrile methine protons and nitrile carbon as well as two diastereoisomers, dhurrin and taxiphyllin (Nahrstedt *et al.*, 1993). Since racemization of cyanogenic glucosides have been known to occur under alkaline conditions (Seigler, 1975), more specific studies are required to determine whether only **47** or both conformers exist without racemization.

Recently, two jasmonoid glucosides (**50** and **51**), a phenylvaleric acid glucoside (**52**), and decenoic acid glucoside (**53**) are also present in the butanol-soluble fraction obtained from methanolic extract of green Perilla leaves, along with phenolics **26** – **29** and their methyl esters, cyanogenic glucosides **47** – **49** and methyl α-**D**-galactopyranoside (Fujita *et al.*, 1996a, b). The structure of **50** was determined to be 5'-β-**D**-glucopyrano-syloxyjasmonic acid, and its absolute configurations at C-1 and C-2 were both assigned to be *R*. The structure of **51** was characterized as 3-β-**D**-glucopyranosyl-3-epi-2-isocucurbic acid. Its absolute configuration at C-3 position was judged to be *R* configuration by applying the glucosylation shifts in ^{13}C-NMR spectroscopy. In a similar manner, **52** was elucidated to be 3-β-**D**-glucopyranosyloxy-5-phenylvaleric acid, and its absolute configuration at C-3 position assigned to be *R*. The structure of **53** was determined to be 5-β-**D**-glucopyranosyloxy-(*Z*)-7-decenoic acid.

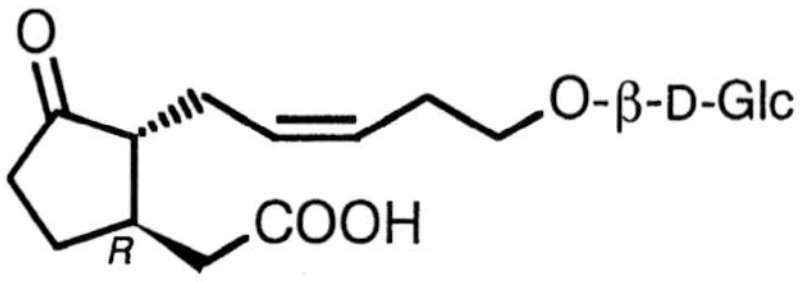

Figure 8   Jasmonoids, phenylvaleric acid, and other glucosides from *P. frutescens*.

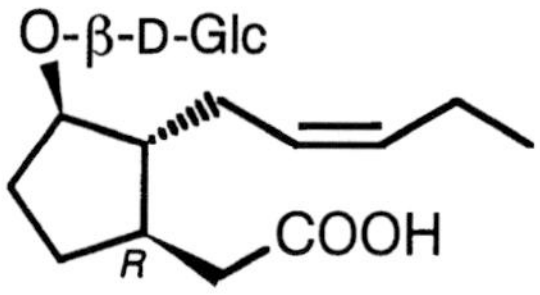

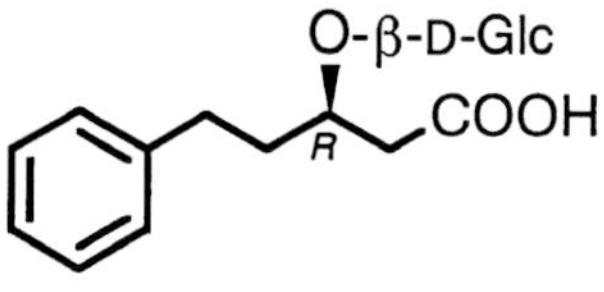

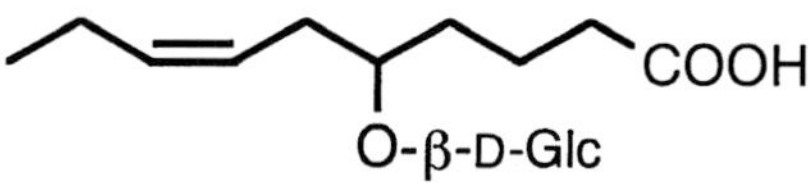

Some other glycosides have been isolated from the same plant extract, and the determination of their structures and the absolute configuration at the C-5 position of **53** is now in progress.

Other Constituents

The composition of fatty acids in Perilla seed oil is so characteristic among various edible oils that it is comprised of highly unsaturated fatty acids, mainly α-linolenic acid which is about 50 %. The study of fatty acid distribution in Perilla seed lipids have been reported by Noda and Obata (1975). Recently, it has been shown that the Perilla seed oil (*n*-3 family), called "egoma oil" (in Japanese), has beneficial effects as compared to common *n*-6 family oils, and it enhances brain activity and nerve systems, and also suppressed the development of cancer, thrombosis and allergic reaction in the animal experiments (Hirano *et al.*, 1991). Palmitic, linoleic, and isooleic acids are also present as lipids or free fatty acids in the chloroform-methanol extract of green Perilla leaves (Nakatsu *et al.*, 1984).

BIOLOGICAL ACTIVITIES OF CONSTITUENTS OF PERILLA

A large number of studies on the biological and pharmacological activities of the constituents of Perilla have been made, but some of these were carried out on crude substances such as an essential oil or a crude extract of the plant. The main biological activities of Perilla components are summarized here.

Volatile Components in Essential Oils

Commercial Perilla oil is widely utilized as a flavoring as are also Perilla leaves and flower stalks, since it can increase the appetite. Therefore, it is not only used in many Japanese processed foods but has also been reported to have some functional applications for masking fishy odor and as an antimicrobial agent and so forth. Using sensory test on the odor of sardines boiled with its major compound perillaldehyde (**1**), it has been shown that **1** contributed to masking the odor specific to boiled sardines (Kasahara and Nishibori, 1988). The antifungal activity of **1** has been investigated, and **1** showed growth inhibition against dermatophytic fungus, e.g., *Trichophyton violaceum* (Kurita and Koike, 1979). Its activity was synergistically increased by adding salt in an agar media (Kurita *et al.*,1981). Honda *et al.* (1984) have also reported that **1** and citral (**9**) were found to show a synergistic inhibitory effect on dermatophytic fungal growth. In addition, the compound **1** has unique biological activities such as a sedative and acaricidal. It has been reported that the oral administration of **1** (100 mg/kg) as well as that of an aqueous extract (4.0 g/kg body wt.) of Perilla leaves prolonged sleep induced by hexobarbital-Na in mice (Sugaya *et al.*, 1981). Sedative activity of the combined effect of **1** and stigmasterol (**21**) has further been discovered by Honda *et al.* (1986) and a significant prolongation of sleep was observed when **1** (2.5 mg/kg) and **21** (5.0 mg/kg) were combined. Insecticidal activity against a tick, namely, acaricide, of **1** has also been reported (Morimoto *et al.*, 1989).

Furthermore, the biological activities of the other components in the Perilla essential oils have been studied. For example, perillaketone (**15**) was isolated from a Perilla essential oil as an active principle of intestinal propulsion in mice (Koezuka *et al.*, 1985b), and **15** was found to be a potent and lung-selective pulmonary toxicant in mice (Wilson *et al.*, 1977; Garst and Wilson, 1984). Dillapiole (**12**) was isolated from a Perilla essential oil as an active principle for prolonging hexobarbital-induced sleep in mice (ED_{50}=1.57 mg/kg) (Honda *et al.*, 1988).

Recently, anti-tumor activity of (-)-perillyl alcohol (**3**) has been investigated. Stark *et al.* (1995) have reported the chemotherapeutic effects of **3** on pancreatic cancer. Gould *et al.* have described that **3** and (+)-limonene inhibited the growth of mammary tumors and induced apoptosis in rat liver tumors (Ren *et al.*, 1994; Shi *et al.*, 1995; Mills *et al.*, 1995). They have also been shown to inhibit protein prenylation and cell proliferation (Gelb *et al.*, 1994; Crowell *et al.*, 1994).

Triterpenoids

Triterpenoids ursolic acid (**23**) and oleanolic acid (**24**) were isolated from Perilla leaves as anti-tumor promoting active substances (Koshimizu *et al.*, 1988, 1991). They were found to show an inhibitory effect on epstein-barr virus activation induced by a 12-O-tetradecanoylphorbol-13-acetate (TPA) (Ohigashi *et al.*, 1986). The treatment of **23** alone (41 nmol) or both **23** and **24** (41 nmol of each) when applied continuously before each TPA-treatment (4.1 nmol) remarkably delayed the formation of papillomas in mouse skin as compared with the control only with TPA (Tokuda *et al.*, 1986). Moreover, tormentic acid (**25**), which was obtained from other several plants (Takahashi *et al.*, 1974; Numata *et al.*, 1989), was found to show a hypoglycemic activity on rats (Villar *et al.*, 1986). They investigated the effect of **25** on its activity by using normoglycemic, hyperglycemic, and streptozotocin diabetic rats as compared with glibenclamide, and then showed that these results suggested that **25** acts by increasing insulin secretion from the islets of Langerhans (Ivorra *et al.*, 1988). Antimicrobial activity of **25**, and also **23** and **24**, against *Streptococcus mutans* has been reported by Isobe *et al.* (1989).

Phenolics and Cinnamates

Phenolic compounds are generally known to have an antioxidative activity. So in common with other members of the Labiatae family, the Perilla extract shows an antioxidative activity because it includes a rich mixture of phenolics, e.g., tocopherols, rosmarinic acid (**26**) and caffeic acid (**27**) in it (Su *et al.* 1986). Three caffeates, methyl caffeate (**32**), vinyl caffeate (**33**), and *p*-menth-8-en-7-yl caffeate (**34**) have also been isolated as antioxidative compounds (Matsumoto *et al.*, 1995). Furthermore, two caffeates, 2-(3,4-dihydroxyphenyl)ethenyl caffeate (**30**) and 2-(3,5-dihydroxyphenyl)ethenyl caffeate (**31**) have been isolated as inhibitors of xanthine oxidase (Nakanishi *et al.*, 1990). Recently, some phenolics **26**, **27**, and protocatechuic aldehyde (**29**) were found to show an inhibitory effect on tumor necrosis factor production (Kosuna *et al.*, 1995). In addition, **26** was isolated from Perilla leaves as a tannic active substance (Okuda *et al.*, 1986), while it was isolated as an active substance on an anti-histamine release activity ($IC_{50}=18\mu M$) against compound 48/80 from *Ehretia philippinensis* (Simpol *et al.*, 1994). The anti-inflammatory activity of **26** determined by the inhibition of malondialdehyde formation in human platelets (Gracza *et al.* 1985), and the effect of **26** as a skin conditioner (Fukushima *et al.*, 1988) have been reported.

Flavonoids and Anthocyanins

Flavonoids and anthocyanins are generally known to have an antioxidative activity as well as phenolics. Additionally, there are many reports on several enzymatic inhibitory effects associated with flavonoids obtained from a number of plants and foods. Anthocyanins obtained from purple Perilla are also important as natural colorants which are relatively stable among other natural colorants. The commercial production of anthocyanin pigments by using cell cultures induced from Perilla has been investigated (Koda *et al.*, 1992).

Glycosides

Monoterpene glucosides, perillosides A (**39**) and C (**41**) obtained from green Perilla leaves, were found to be inhibitors of aldose reductase, and the effect of related glucosides and their tetraacetates on aldose reductase has also been elucidated (Fujita *et al.*, 1995).

Aldose reductase (EC 1.1.1.21, abbr. AR) is a key enzyme of the polyol pathway, which catalyzes the reduction of hexoses to sugar alcohols (Gabbay, 1973). It is expected that AR inhibitors will play an important role in the management of diabetic complications such as cataract, retinopathy, neuropathy, and nephropathy (Kinoshita *et al.*, 1981; Sakamoto and Hotta 1983). We observed the inhibitory effects of glucosides derived from Perilla and related monoterpene glucosides (Figure 9), which were prepared by modified Koenigs-Knorr method (Paulsen *et al.*, 1985; Schwab *et al.*, 1990), against rat lens AR (RLAR) and human recombinant AR (HRAR) in order to elucidate the relationship

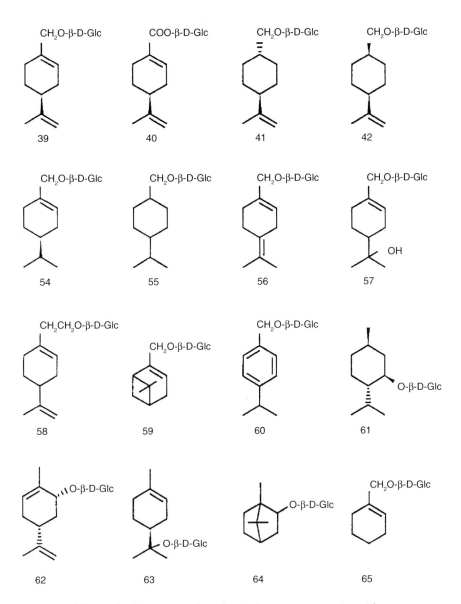

Figure 9   Structures of synthesized monoterpene glucosides

between the structure and inhibitory activity. AR assays using partially purified RLAR and purchased HRAR were conducted according to the method by Hayman *et al.* (1965).

The inhibitory effects of monoterpene glucosides including perillosides on RLAR are shown in Table 2. Perilloside A (**39**) and C (**41**), and their homologues (-)-phellandryl β-**D**-glucopyranoside (**54**) and 1, 4(8)-*p*-menthadien-7-yl β-**D**-glucopyranoside (**56**), and another type, (-)-menthyl β-**D**-glucopyranoside (**61**) were quite inhibitory to RLAR,

Table 2   Inhibitory effect of glucosides and their tetraacetates on RLAR

Glucoside	Glc (%)		Glc (OAc)$_4$ (%)	
	0.1 mM	0.01 mM	0.1 mM	0.01 mM
Perilloside A (39)	54.5	6.7	87.1	39.6
Perilloside B (40)	14.4	11.1	30.7	25.5
Perilloside C (41)	46.4	26.1	67.0	30.7
Perilloside D (42)	28.2	14.8	45.2	13.7
(-)-Phellandryl β-D-Glc (54)	43.6	15.7	65.4	10.5
p-Menthan-7-yl β-D-Glc (55)	26.8	13.9	62.3	40.7
1,4(8)-p-Menthadien-7-yl β-D-Glc (56)	41.9	32.9	42.3	9.1
1-p-Menthene-7,8-diol 7-β-D-Glc (57)	14.8	3.8	26.3	9.5
Homoperillyl β-D-Glc (58)	35.6	17.4	16.2	0
Myrtenyl β-D-Glc (59)	16.3	8.7	37.7	10.5
Cuminyl β-D-Glc (60)	9.5	5.5	-	-
(-)-Menthyl β-D-Glc (61)	38.7	11.6	-	4.2
(-)-cis-Carvyl β-D-Glc (62)	9.1	0	38.5	16.2
(-)-α-Terpinyl β-D-Glc (63)	14.2	0	20.8	17.5
Bornyl β-D-Glc (64)	10.4	8.7	33.5	23.8
1-Cyclohexenylmethyl β-D-Glc (65)	2.0	0	62.3	36.6
α-Type glucoside				
(-)-Perillyl α-D-Glc (66)	20.1	9.1	37.6	7.1
(-)-Phellandryl α-D-Glc (67)	31.7	16.4	65.4	18.7

-: Not measured.

where all of the corresponding aglycones, alcohols and acid were inactive in this assay. Their tetraacetates demonstrated an approximately one order higher activity than the corresponding glucosides. The α-D-type glucosides, (-)-perillyl α-D-glucopyranoside (66) and (-)- phellandryl α-D-glucopyranoside (67) and their tetraacetates had a lower inhibitory activity in each case. In the same manner, the inhibitory effects of the glucosides 39, 41, and 54, and their tetraacetates on HRAR were screened and then their inhibitory tendencies were similar to those of the case of RLAR, although the activities were somewhat weaker at the same concentration. These results may suggest as follows; (1) the p-menthane skeleton having a glucosyloxy moiety at the C-7 position, e.g., perilloside A (39) is essential for the appearance of inhibitory action; (2) double bonds in the p-menthane skeleton increased the activity; (3) equatorial substituents of 39, 41, and 54, when the compounds have a stable chair form, were better than axial ones as in 42 and 55; (4) acetylation of the glucosides resulted in a one order higher activity than the original glucosides; (5) the glucosidic β-linkage with aglycone was preferred to the α-linkage, the former favoring a planar structure; (6) an ether-type glucoside was better than an ester-type one.

Kinetic studies were conducted on the compound 39 and tetraacetylperilloside A in order to determine the type of inhibition and inhibition constant (Ki). The Ki values of 39 for the substrates of glyceraldehyde and NADPH, were calculated at 1.4×10^{-4} M and 4.3×10^{-4} M from the Lineweaver-Burk plots, respectively. The apparent type of enzyme inhibition by **39** was competitive with respect to glyceraldehyde, while non-competitive to NADPH. On the other hand, the Ki values of tetraacetylperilloside A for the substrates of glyceraldehyde and NADPH, were calculated at 2.5×10^{-5} M and 4.0×10^{-5} M, respectively. Its type of inhibition was non-competitive with the both glyceraldehyde and NADPH. In addition, kinetic study of **39** and its tetraacetate with HRAR was conducted, and their type of inhibition were same as those with RLAR. In this case, the Ki values of **39** and tetraacetylperilloside A exhibited 4.3×10^{-4} M and 1.1×10^{-4} M for glyceraldehyde, respectively.

This was the first report of monoterpene glucosides such as perillosides as AR inhibitors, although many flavonoids, carboxylic acids, and tannins have historically been examined as AR-inhibitory drugs (Okuda *et al.*, 1984; Kador *et al.*, 1983, 1985; Raskin and Rosenstock, 1987). The inhibitor binding should result from a combination of hydrophobic binding and a reversible charge-transfer reaction. In this experiment, the introduction of acetyl groups as lipophilic substituents into monoterpene glucosides increased the magnitude of the inhibition of AR. Furthermore, the β-form of monoterpene glucosides showed a higher inhibition than the α-form. These results indicate that ARs have stereochemical requirements for binding, as well as lipophilic binding regions and a charge-transfer pocket. More specific studies are required to elucidate the stereochemical requirements for binding to ARs. However, the present results should contribute to the design of more effective AR inhibitors.

In recent years, we have become interested in biological activities such as the plant growth regulatory activity of jasmonoids (Yamane, 1994). The jasmonoid glucosides **50** and **51** obtained from the green Perilla were stereoisomers of β-**D**-glucopyranosyltuberonic acid (= 5'-β-**D**-glucopyranosyloxy-2-epijasmonic acid) and (1R)-3-β-**D**-glucopyranosylcucurbic acid, respectively. The former have been isolated as a tuber-inducing substance from potato leaves (Yoshihara *et al.*, 1989), and the later have been isolated as a plant growth inhibitor from *Cucurbita pepo* L. (Fukui *et al.*, 1977). Although these jasmonoid glucosides have been a little reported, a number of jasmonoids have been isolated from plants and microorganisms as a plant growth regulator. The related glucosides **52** and **53** have also been obtained from Perilla leaves. Consequently, these glucosides **50–53** from Perilla are expected to play an important role in plant as well as other jasmonoids.

CONCLUSION

There are a large number of studies on the constituents of Perilla, since the plant has been used for various purposes, e.g., as a garnish, flavoring, and natural colorant. These results suggest that the constituents of Perilla have various beneficial properties compared with other edible plants. Perilla essential oil is not only widely utilized as a flavoring but also used in many processed foods to increase appetite. Therefore, many volatile compounds in Perilla oil have been investigated, and some of their components have been reported to show an antimicrobial, sedative, and anti-tumor activities. A number of non-volatile compounds, on the other hand, terpenoids, phenolics, flavonoids, and glycosides have been isolated from Perilla extract, and have also been described to have some biological activities, for example, an anti-tumor promoting, hypoglycemic, antioxidative, anti-inflammatory, and aldose reductase inhibitory activities. Recently, the effect of the Perilla water extract on anti-allergic activity has been noted, as has its effectiveness against some allergic symptoms caused by pollen, foods, etc., although the active principle has not been fully identified. If the beneficial properties ascribed to Perilla are made clearer by biological and chemical evidence, the plant will arouse even greater interest.

REFERENCES

Arctander, S. (1969) *Perfume and Flavor Chemicals*, vols. I and II, Montclair, New Jersey, pp. 937.

Aritomi, M. (1982) Chemical studies on edible plant. I. Phenolic constituents of *Perilla frutescens*. *Kaseigaku Zasshi*, **33**, 353–359.

Aritomi, M., Kumori, T. and Kawasaki, T. (1985) Cyanogenic glycosides in leaves of *Perilla frutescens* var. *acuta*. *Phytochemistry*, **24**, 2438–2439.

Aritomi, M. (1988) Chemical studies on the constituents of edible plants. VI. Cyanogenecity and practical use of *Perilla frutescens* var. *acuta* as a food colorant. *Nippon Kasei Gakkaishi*, **39**, 817–822.

Crowell, P.L., Ren, Z., Lin, S., Vedejs, E. and Gould, M.N. (1994) Structure-activity relationships among monoterpene inhibitors of protein isoprenylation and cell proliferation. *Biochem. Pharmacol.*, **47**, 1405–1415.

Fujita, T. and Nakayama, M. (1992) Perilloside A, a monoterpene glucoside from *Perilla frutescens*. *Phytochemistry*, **31**, 3265–3267.

Fujita, T. and Nakayama, M. (1993) Monoterpene glucosides and other constituents from *Perilla frutescens*. *Phytochemistry*, **34**, 1545–1548.

Fujita, T., Funayoshi, A. and Nakayama, M. (1994) A phenylpropanoid glucoside from *Perilla frutescens*. *Phytochemistry*, **37**, 543–546.

Fujita, T., Ohira, K., Miyatake, K., Nakano, Y. and Nakayama, M. (1995) Inhibitory effect of perillosides A and C, and related monoterpene glucosides on aldose reductase and their structure-activity relationships. *Chem. Pharm. Bull.*, **43**, 920–926.

Fujita, T., Terato, K. and Nakayama, M. (1996a) Two jasmonoid glucosides and a phenylvaleric acid glucoside from *Perilla frutescens*. *Biosci. Biotech. Biochem.*, **60**, 732–735.

Fujita, T., Terato, K. and Nakayama, M. (1996b) Isolation and structure elucidation of the glucosidic constituents from *Perilla frutescens*. *Koen Yoshisyu – Nippon Nogeikagaku Kaishi*, **70**, pp. 238 (in Japanese).

Fujita, Y., Mizohata, H. and Iwamura, J. (1966) Essential oil of *Perilla frutescens*. IX. *Nippon Kagaku Kaishi*, **87**, 1361–1363 (in Japanese).

Fujita, Y., Fujita, S. and Hayama, Y. (1970a) *trans*-Shisool (*trans*-8-*p*-menthen-7-ol) isolated from the essential oils of *Perilla acuta* Nakai and *P. acuta* f. *viridis* Nakai. *Bull. Chem. Soc. Jpn.*, **43**, 2637–2638.

Fujita, Y., Fujita, S. and Hayama, Y. (1970b) Miscellaneous contributions to the essential oils of the plants from various territories. XXIV. Essential oils of *Perilla acuta* (Thunb.) Nakai and *P. acuta* f. *viridis* (Makino) Nakai. *Nippon Nogeikagaku Kaishi*, **44**, 428-432 (in Japanese).

Fukui, H., Koshimizu, K., Yamazaki, Y, and Usuda, S. (1977) Structures of plant growth inhibitors in seeds of *Cucurbita pepo* L. *Agric. Biol. Chem.*, **41**, 189–194.

Fukushima, M., Yagisawa, T., Kinoshita, K. and Sano, H. (1988) Cosmetics containing rosmarinic acid and/or its salts as skin conditioners. Jpn. Kokai Tokkyo Koho JP63-162611 (in Japanese)(*Chem. Abstr.*, **110**: P218817h).

Gabbay, K.H. (1973) The sorbitol pathway and the complications of diabetes. *N. Engl. J. Med.*, **288**, 831–836.

Garst, J.E. and Wilson, B.J. (1984) Synthesis and analysis of various 3-furyl ketones from *Perilla frutescens*. *J. Agric. Food Chem.*, **32**, 1083–1087.

Gelb, M.H., Tamanoi, F., Yokoyama, K., Ghomashchi, F., Esson, K. and Gould, M.N. (1995) The inhibition of protein prenyltransferases by oxygenated metabolites of limonene and perillyl alcohol. *Cancer Lett.*, **91**, 169–175.

Gracza, L., Koch, H. and Loeffler, E. (1985) Biochemical-pharmacological investigations of medicinal agents of plant origin. I. Isolation of rosmarinic acid from *Symphytum officinale* L. and its anti-inflammatory activity in an *in vitro* model. *Arch. Pharm.* (Weinheim), **318**, 1090–1095.

Hayman, S. and Kinoshita, J.H. (1965) Isolation and properties of lens aldose reductase. *J. Biol. Chem.*, **240**, 877–882.

Hirano, J., Isoda, Y. and Nishizawa, Y. (1991) Utilization of *n*-3 plant oils perilla and flaxseed oils. *Yukagaku*, **40**, 942–950 (in Japanese).

Honda, G., Koga, K., Koezuka, Y. and Tabata, M. (1984) Antidermatophytic compounds of *Perilla frutescens* Britton var. *crispa* Decne. *Shoyakugaku Zasshi*, **38**, 127–130 (in Japanese).

Honda, G., Koezuka, Y., Kamisako, W. and Tabata, M. (1986) Isolation of sedative principles from *Perilla frutescens*. *Chem. Pharm. Bull.*, **34**, 1672–1677.

Honda, G., Koezuka, Y. and Tabata, M. (1988) Isolation of dillapiole from a chemotype *Perilla frutescens* as an active principle for prolonging hexobarbital-induced sleep. *Chem. Pharm. Bull.*, **36**, 3153–3155.

Honda, G. (1994) Genetics of essential oil components of *Perilla frutescens*. *Farumashia*, **30**, 486–490 (in Japanese).

Ina, K. and Ogura, I. (1970) Studies on the components of perilla essential oil. I. Neutral essential oil. *Nippon Nogeikagaku Kaishi*, **44**, 209–212 (in Japanese).

Ina, K. and Suzuki, I. (1971) Studies on the components of perilla essential oil. II. Furan derivatives in neutral essential oil. *Nippon Nogeikagaku Kaishi*, **45**, 113–117 (in Japanese).

Ishikura, N. (1981) Anthocyanins and flavones in leaves and seeds of *Perilla* plant. *Agric. Biol. Chem.*, **45**, 1855–1860.

Isobe, T., Noda, Y., Ohsaki, A., Sakanaka, S., Kim, M. and Taniguchi, M. (1989) Studies on the constituents of *Leucoseptrum stellipillum*. *Yakugaku Zasshi*, **109**, 175–178 (in Japanese).

Ito, H. (1966) Studies on the Folium Perillae. IV. *Perilla* spp. containing myristicin and dillapiole as the main components of the essential oils. *Shoyakugaku Zasshi*, **20**, 73–75 (in Japanese).

Ito, H. (1968) Studies on the Folium Perillae. V. *Perilla* spp. containing elemicin as the main components of the essential oils. *Shoyakugaku Zasshi*, **22**, 151–152 (in Japanese).

Ito, H. (1970) Studies on Folium Perillae. VI. Constituent of essential oils and evaluation of genus *Perilla*. *Yakugaku Zasshi*, **90**, 883–892 (in Japanese).

Ivorra, M.D., Paya, M. and Villar, A. (1988) Hypoglycemic and insulin release effects of tormentic acid: A new hypoglycemic natural product. *Planta Medica*, **1988**, 282–286.

Kador, P.F. and Sharpless, N.E. (1983) Pharmacophor requirements of the aldose reductase inhibitor site. *Mol. Pharmacol.*, **24**, 521–531.

Kador, P.F., Kinoshita, J.H. and Sharpless, N.E. (1985) Aldose reductase inhibitors: A potential new class of agents for the pharmacological control of certain diabetic complications. *J. Med. Chem.*, **28**, 841–849.

Kameoka, H. and Nishikawa, K. (1976) The composition of the essential oil from *Perilla frutescens* L. Brit. var. *acuta* Thunb. Kudo and *Perilla frutescens* L. Brit. var. *acuta* Thunb. Kudo f. *discolor* Makino. *Nippon Nogeikagaku Kaishi*, **50**, 345-349 (in Japanese).

Kasahara, K. and Nishibori, K. (1988) Suppressing effect of perilla for odor of sardine. *Nippon Suisan Gakkaishi*, **54**, 315–317 (in Japanese).

Kinoshita, J.H., Kador, P. and Catiles, M. (1981) Aldose reductase in diabetic cataracts. *J. Am. Med. Assoc.*, **246**, 257–261.

Koda, T., Ichi, T., Yoshimitu, M., Nihongi, Y. and Sekiya, J. (1992) Production of perilla pigment in cell cultures of *Perilla frutescens*. *Nippon Shokuhin Kogyo Gakkaishi*, **39**, 839–844 (in Japanese).

Koezuka, Y., Honda, G. and Tabata, M. (1984) Essential oil types of the local varieties and their F_1 hybrids of *Perilla frutescens*. *Shoyakugaku Zasshi*, **38**, 238–242 (in Japanese).

Koezuka, Y., Honda, G., Sakamoto, S. and Tabata, M. (1985a) Genetic control of anthocyanin production in *Perilla frutescens*. *Shoyakugaku Zasshi*, **39**, 228–231.

Koezuka, Y., Honda, G. and Tabata, M. (1985b) An intestinal propulsion promoting substance from *Perilla frutescens* and its mechanism of action. *Planta Medica*, **1985**, 480–482.

Koezuka, Y., Honda, G. and Tabata, M. (1986a) Genetic control of the chemical composition of volatile oils in *Perilla frutescens*. *Phytochemistry*, **25**, 859–863.

Koezuka, Y., Honda, G. and Tabata, M. (1986b) Genetic control of phenylpropanoids in *Perilla frutescens*. *Phytochemistry*, **25**, 2085–2087.

Koezuka, Y., Honda, G. and Tabata, M. (1986c) Genetic control of isoegomaketone formation in *Perilla frutescens*. *Phytochemistry*, **25**, 2656–2657.

Kondo, T., Tamura, H., Yoshida, K. and Goto, T. (1989) Structure of malonylshisonin, a genuine pigment in purple leaves of *Perilla ocimoides* L. var. *crispa* Benth. *Agric. Biol. Chem.*, **53**, 797–800.

Koshimizu, K., Ohigashi, H., Tokuda, H., Kondo, A. and Yamaguchi, K. (1988) Screening of edible plants against possible anti-tumor promoting activity. *Cancer Letters*, **39**, 247–257.

Koshimizu, K. (1991) Anti-tumor promoting effect of the constituents in foods. *Kagaku to Seibutsu*, **29**, 598–603 (in Japanese).

Kosuna, K., Shirai, J. and Kosaka, H. (1995) Anti-inflammatory effect of the constituents from perilla extract. *Fragrance J.*, **1995**, 90–94 (in Japanese).

Kurita, N., Miyaji, M., Kurane, R., Takahara, Y. and Ichimura, K. (1979) Anti-fungal activity and molecular orbital energies of aldehyde compounds from oils of higher plants. *Agric. Biol. Chem.*, **43**, 2365–2371.

Kurita, N. and Koike, S. (1981) Synergistic antimicrobial effect of perilla and NaCl. *Nippon Nogeikagaku Kaishi*, **55**, 43–46 (in Japanese).

Masada, Y. (1975) *Analysis of Essential Oils by Gas Chromatography and Mass Spectrometry*, Hirokawa Publishing Co., Tokyo, pp. 27–30.

Matsumoto, R., Yamaguchi, H., Chiba, K. and Tada, M. (1995) Antioxidative compounds and novel monoterpenoid from *Perilla frutescens*. *Koen Yoshisyu – Koryo, Terupen oyobi Seiyu Kagaku ni Kansuru Toronkai*, 39th, pp. 199–201 (in Japanese).

Mills, J.J., Chari, R.S., Boyer, I.J., Gould, M.N. and Jirtle, R.L. (1995) Induction of apoptosis in liver tumors by the monoterpene perillyl alcohol. *Cancer Res.*, **55**, 979–983.

Misra, L. N. and Husain, A. (1987) The essential oil of *Perilla ocimoides*: A rich source of rosefuran. *Planta Medica*, **53**, 379–380.

Morimoto, Y., Takaoka, K. and Watanabe, T. (1989) Acaricidal effect of essential oils. Jpn. Kokai Tokkyo Koho JP01-19004 (in Japanese).

Mulkens, A. and Kapetanidis, I. (1988) Eugenylglucoside, a new natural phenylpropanoid heteroside from *Melissa officinalis*. *J. Nat. Prod.*, **51**, 496–498.

Nabeta, K., Ohnishi, Y., Hirose, T. and Sugisawa, H. (1983) Monoterpene biosynthesis by callus tissues and suspension cells from *Perilla* species. *Phytochemistry*, **22**, 423–425.

Nabeta, K., Oda, T., Fujimura, T. and Sugisawa, H. (1984) Biosynthesis of cuparene from mevalonic acid-6,6,6-$^2H_3$ by *in vitro* callus culture of *Perilla* sp. *Agric. Biol. Chem.*, **48**, 3141–3143.

Nabeta, K., Kawakita, K., Yada, Y. and Okuyama, H. (1993) Biosynthesis of sesquiterpenes from deuterated mevalonates in *Perilla* callus. *Biosci. Biotech. Biochem.*, **57**, 792–798.

Nahrstedt, A., Lechtenberg, M., Brinker, A., Seigler, D.S. and Hegnauer, R. (1993) 4-Hydroxy-mandelonitrile glucosides, dhurrin in *Suckleya suckleyana* and taxiphyllin in *Girgensohnia oppositiflora* (Chenopodiaceae). *Phytochemistry*, **33**, 847–850.

Nakanishi, T., Nishi, M., Inada, A., Obata, H., Tanabe, N., Abe, S. and Wakashiro, M. (1990) Two new potent inhibitors of xanthine oxidase from leaves of *Perilla frutescens* Britton var. *acuta* Kudo. *Chem. Pharm. Bull.*, **38**, 1772–1774.

Nakatsu, S., Tomita, K., Nakatsuru, I. and Matsuda, K. (1984) On the lipids in vegetables. I. Fatty acid composition of lipids from vegetables. Bull. Fac. Agric. Miyazaki Univ., **31**, 21–32 (in Japanese).

Nishizawa, A., Honda, G. and Tabata, M. (1989) Determination of final steps in biosyntheses of essential oil components in *Perilla frutescens*. *Planta Medica*, **55**, 251–253.

Nishizawa, A., Honda, G. and Tabata, M. (1990a) Characteristic incorporation of fatty acids into lower terpenoids in cotyledons of *Perilla frutescens*. *Chem. Pharm. Bull.*, **38**, 1317–1319.

Nishizawa, A., Honda, G. and Tabata, M. (1990b) Genetic control of perillene accumulation in *Perilla frutescens*. *Phytochemistry*, **29**, 2873–2875.

Noda, M. and Obata, T. (1975) Fatty acid distribution in perilla seed lipids. *Nippon Nogeikagaku Kaishi*, **49**, 251–256 (in Japanese).

Numata, A., Yang, P., Takahashi, C., Fujiki, R., Nabae, M. and Fujita, E. (1989) Cytotoxic triterpenes from a Chinese medicine, Goreishi. *Chem. Pharm. Bull.*, **37**, 648–651.

Ohigashi, H., Takamura, H., Koshimizu, K., Tokuda, H. and Ito, Y. (1986) Search for possible antitumor promoters by inhibition of 12-O-tetradecanoylphorbol-13-acetate-induced epstein-barr virus activation; Ursolic acid and oleanolic acid from an anti-inflammatory chinese medicinal plant, *Glechoma hederaceae* L. *Cancer Letters*, **30**, 143–151.

Okuda, J., Miwa, I., Inagaki, K., Horie, T. and Nakayama, M. (1984) Inhibition of aldose reductase by 3',4'-dihydroxyflavones. *Chem. Pharm. Bull.*, **32**, 767–772.

Okuda, O. (1967) *Koryo Kagaku Souran*, Hirokawa Publishing Co., Tokyo, pp. 342-344 (in Japanese).

Okuda, T., Hatano, T., Agata, I. and Nishibe, S. (1986) The components of tannic activities in Labiatae plants. I. Rosmarinic acid from Labiatae plants in Japan. *Yakugaku Zasshi*, **106**, 1108–1111 (in Japanese).

Part, H.S., Kim, J. G. and Cho, M.J. (1982) Chemical compositions of *Perilla frutescens* Britton var. *crispa* Decaisne cultivated in different area of Korea. II. Sterol compositions. *Hanguk Nonghwa Hakhoe Chi*, **25**, 14–20 (*Chem. Abstr.*, **97**:78761w) (in Korean).

Paulsen, H., Le-Nguyen, B., Sinnwell, V., Heemann, V. and Seehofer, F. (1985) Synthese von glycosiden von mono-, sesqui- und diterpenalkoholen. *Liebigs Ann. Chem.*, **1985**, 1513–1536.

Raskin, P. and Rosenstock, J. (1987) Aldose reductase inhibitors and diabetic complications. *Am. J. Med.*, **83**, 298–306.

Ren, Z. and Gould, M.N. (1994) Inhibition of ubiquinone and cholesterol synthesis by the monoterpene perillyl alcohol. *Cancer Lett.*, **76**, 185–190.

Sakai, T. and Hirose, Y. (1969) Farnesenes isolated from the volatile oil of *Perilla frutescens* f. *viridis* Makino. *Bull. Chem. Soc. Jpn.*, **42**, 3615.

Sakamoto, N. and Hotta, N. (1983) Inhibitors of aldose reductase and their clinical applications. *Farumashia*, **19**, 43–47 (in Japanese).

Schwab, W. and Schreier, P. (1988) Aryl β-**D**-glucosides from *Carica papaya* fruit. *Phytochemistry*, **27**, 1813–1816.

Schwab, W., Scheller, G. and Schreier, P. (1990) Glycosidically bound aroma components from sour cherry. *Phytochemistry*, **29**, 607–612.

Schwarzmaier, U. (1976) Notiz zur Konfigurationsbestimmung bei freien mandelsaurenitril-glycosiden. *Chem. Ber.*, **109**, 3250–3251.

Seigler, D.S. (1975) Isolation and characterization of naturally occurring cyanogenic compounds. *Phytochemistry*, **14**, 9–29.

Shi, W. and Gould, M.N. (1995) Induction of differentiation in neuro-2A cells by the monoterpene perillyl alcohol. *Cancer Lett.*, **95**, 1–6.

Simpol, L.R., Otsuka, H., Ohtani, K., Kasai, R. and Yamasaki, K. (1994) Nitrile glucosides and rosmarinic acid, the histamine inhibitor from *Ehretia philippinensis. Phytochemistry*, **36**, 91–95.

Stark, M.J., Burke, Y.D., McKinzie, J.H., Ayoubi, A.S. and Crowell, P.L. (1995) Chemotherapy of pancreatic cancer with the monoterpene perillyl alcohol. *Cancer Lett.*, **96**, 15–21.

Su, J.-D., Osawa, T. and Namiki, M. (1986) Screening for antioxidative activity of crude drugs. *Agric. Biol. Chem.*, **50**, 199–203.

Suga, T., Hirata, T., Aoki, T. and Shishibori, T. (1986) Interconversion and cyclization of acyclic allylic pyrophosphates in the biosynthesis of cyclic monoterpenoids in higher plants. *Phytochemistry*, **25**, 2769–2775.

Sugaya, A., Tsuda, T. and Obuchi, T. (1981) Pharmacological studies of Perillae Herba. I. Neuropharmacological action of water extract and perillaldehyde. *Yakugaku Zasshi*, **101**, 642–648.

Takagi, S. (1985) Determination of green leaf carotenoids by HPLC. *Agric. Biol. Chem.*, **49**, 1211–1213.

Takahashi, K., Kawaguchi, S., Nishimura, K., Kubota, K., Tanabe, Y. and Takani, M. (1974) Studies on constituents of medicinal plants. XIII. Constituents of the pericarps of the capsules of *Euscaphis japonica* Pax. (I). *Chem. Pharm. Bull.*, **22**, 650–653.

Tamura, H., Fujiwara, M. and Sugisawa, H. (1989) Production of phenylpropanoids from cultured callus tissue of the leaves of akachirimen-shiso (*Perilla* sp.). *Agric. Biol. Chem.*, **53**, 1971–1973.

Tokuda, H., Ohigashi, H., Koshimizu, K. and Ito, Y. (1986) Inhibitory effects of ursolic and oleanolic acid on skin tumor promotion by 12-O-tetradecanoylphorbol-13-acetate. *Cancer Letters*, **33**, 279–285.

Tomita, Y., Arata, M. and Ikeshiro, Y. (1985) Biosynthesis of ursolic acid in cell cultures of *Perilla frutescens* Britt. var. *acuta* Kudo: Mechanism of D- and E-ring formation. *J. Chem. Soc., Chem. Commun.*, **1985**, 1087–1088.

Tomita, Y. and Ikeshiro, Y. (1994) Biosynthesis of ursolic acid in cell cultures of *Perilla frutescens. Phytochemistry*, **35**, 121–123.

Villar, A., Paya, M., Hortiguela, M.D. and Cortes, D. (1986) Tormentic acid, a new hypoglycemic agent from *Poterium ancistroides. Planta Medica*, **1986**, 43–45.

Wilson, B.J., Garst, J.E., Linnabary, R.D. and Channell, R.B. (1977) Perilla Ketone: A potent lung toxin from the mint plant, *Perilla frutescens* Britton. *Science*, **197**, 573–574.

Yamane, H. (1994) *Plant Hormone Handbook*, Takahashi, N. and Masuda, Y., Eds., Baifukan, Tokyo, pp. 279–292 (in Japanese).

Yoshida, K., Kondo, T., Kameda, K. and Goto, T. (1990) Structure of anthocyanins isolated from purple leaves of *Perilla ocimoides* L. var. *crispa* Benth and their isomerization by irradiation of light. *Agric. Biol. Chem.*, **54**, 1745–1751.

Yoshida, K., Kameda, K. and Kondo, T. (1993) Diglucuronoflavones from purple leaves of *Perilla ocimoides. Phytochemistry*, **33**, 917–919.

Yoshihara, T., Omer, E.A., Koshino, H., Sakamura, S., Kikuta, Y. and Koda, Y. (1989) Structure of a tuber-inducing stimulus from potato leaves (*Solanum tuberosum* L.). *Agric. Biol. Chem.*, **53**, 2835–2837.

Yuba, A., Honda, G., Mizukoshi, T. and Tabata, M. (1992) Organ-specific expression of a genetic factor inducing monoterpene synthesis in the calyx of *Perilla frutescens. Shoyakugaku Zasshi*, **46**, 257–260.

11. CHEMOTYPES AND PHARMACOLOGICAL ACTIVITIES OF PERILLA

MAMORU TABATA

Sakuragaoka W 8-17-5, Sanyo-cho, Okayama 709-08, Japan
Professor Emeritus, Kyoto University, Yoshida, Kyoto, Japan

CHEMOTYPES OF PERILLA

Variation in Essential Oil Composition

Perilla frutescens Britt. has been used as a popular herb in the East primarily because of a delicate aroma from the essential oil, which has been proved to be biosynthesized and accumulated mainly in the peltate glandular trichome of the leaf (Nishizawa *et al.*, 1992a). It is known that the agreeable aromatic smell largely depends on the presence of a cyclic monoterpenoid, perillaldehyde (4-isopropenyl-1-cyclohexene-1-carboxaldehyde), which occupies a large part of the essential oil. However, there are chemical varieties of this species which do not have perillaldehyde but either a monoterpenoid with a disagreeable smell or a phenylpropanoid as the main oil component, although such varieties should not be used as a traditional medicine according to the old Chinese herbals.

It is therefore necessary not only for medical treatment but also for the pharmacological studies of Perilla to distinguish the perillaldehyde type from the other chemotypes by their chemical compositions of essential oils, since they can not be distinguished by their morphological characters. It is also important to clarify the genetic basis of chemical variations for the purpose of breeding Perilla plants to establish a superior variety showing genetic stability and pharmacologically desirable chemical composition.

On the basis of the main components of the essential oils, Ito (1970) had classified chemical varieties of Perilla into four groups: perillaldehyde (PA), furylketone (FK), phenylpropanoid (PP), and citral (C) types. Recently, Koezuka *et al.* (1984) carried out GC analysis of the essential oils from the leaves of 215 lines of Perilla that had been collected from various parts of Japan to be cultivated under the same field conditions in Kyoto. The results showed that these lines could be classified into five distinct chemotypes according to differences in major oil components: PA (perillaldehyde), EK (elsholtziaketone), PK (perillaketone or isoegomaketone), C (citral), and PP (one or more of specific phenylpropanoids: myristicin, dillapiole, and elemicin). In addition to these five groups, a new chemotype, PL, which contains perillene as a major component of its essential oil was found by Nishizawa *et al.* (1990). The essential oil in the leaves of the PA type consisted of perillaldehyde (71.0%), limonene (9.3%) and caryophyllene (5.8%), a sesquiterpene, whereas that of the PP type contained myristicin (52.7%) and caryophyllene (38.2%) in place of monoterpenoids (Nishizawa *et al.*, 1992).

All these chemotypes proved to be genetically stable, showing no segregation for the chemical composition in the self-pollinated progenies. To investigate the genetic

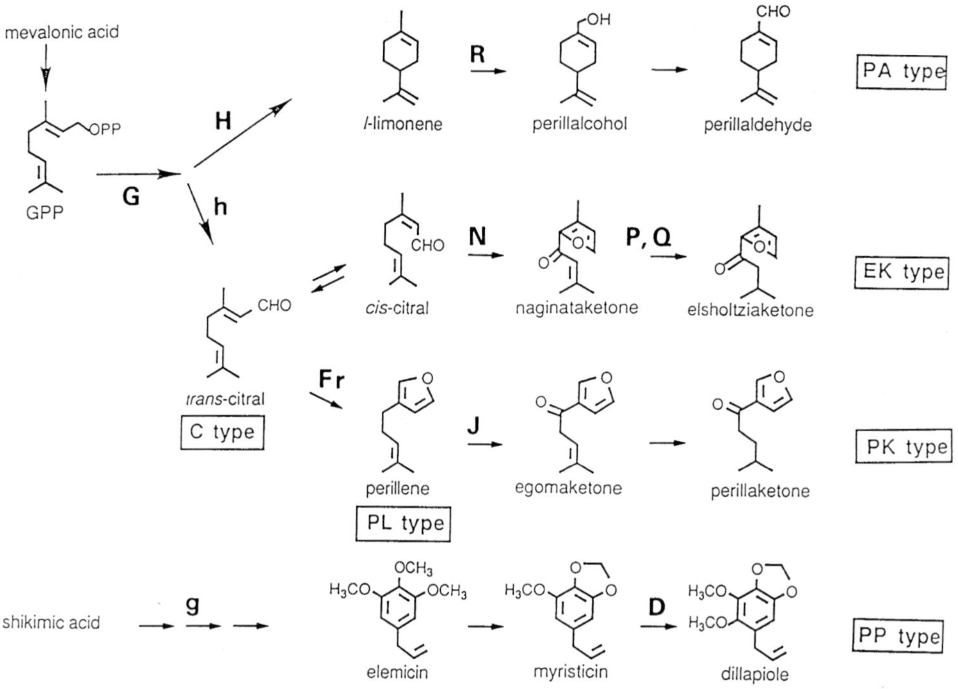

Figure 1   Proposed biosynthetic pathways of essential oil components of *Perilla frutescens*. Symbols in capital letters represent dominant genes controlling possible reaction steps. Symbols in small letters represent recessive alleles. The dominant genes are required for promoting the respective biochemical reactions, except for the dominant inhibitor gene R that blocks the conversion of *l*-limonene into perillalcohol to cause an accumulation of the former in the leaves.

mechanisms for the chemical differences in essential oils, the plants of different chemotypes were artificially intercrossed to obtain the F_1 hybrids and the F_2 progenies, which were then examined individually for the essential oils of their leaves (Koezuka et al., 1986a-c; Nishizawa et al.,1989, 1990, 1991). The results of the genetic experiments have clearly demonstrated that the chemical composition of the essential oils is controlled by several independent genes and that the chemotypes are determined by the genotypes with regard to these genes (Figure 1).

Figure 1 illustrates the roles of these genes in relation to the biosynthetic pathways leading to different final products. The most important role is played by G, a dominant gene which is essential for initiating the biosynthesis of all the monoterpenoids from the mevalonate pathway in Perilla plants. In the absence of the key dominant gene G, namely, in any plants homozygous for the recessive allele g, no monoterpenoids but unusual phenylpropanoids (myristicin, dillapiole, or elemicin) are produced from the shikimic acid pathway. Furthermore, in the presence of the two dominant genes, G and H, the volatile oil of the PA type containing mostly perillaldehyde and l-limonene is accumulated. This suggests that the dominant gene H is required for the formation of cyclic monoterpenes. In addition, another dominant gene, R, has been found to be inhibitory to the conversion of limonene into perillalcohol (Yuba *et al.*, 1995).

In the absence of the dominant gene H, both GGhh and Gghh plants produce acyclic monoterpenoids of any one of EK, PK, PL, and C types, depending on genotypes for such genes as N, P, Q, Fr, and J, which control the succeeding reaction steps after the formation of t-citral. Thus the chemical variety of the PA type, which is considered to be suitable for a crude drug as well as for a spice, proved to possess the genotype GGHHrr, whereas the genotype of the PK chemotype that gives rise to perillaketone, a substance toxic to the animals, was estimated to be GGhhFrFrJJ.

Nishizawa *et al.* (1992b) recently have studied the relationship between the two genes, G and H, and the two key enzymes involved in monoterpenoid biosynthesis, *viz.* the specific geranylpyrophosphate (GPP) synthase that catalyzes the synthesis of GPP from isopentenylpyrophosphate and dimethylallylpyrophosphate and the limonene synthase involved in the cyclization of GPP. The results of the enzyme assay showed that GPP synthase activity was present in all of the chemotypes (PA, EK, and PK) including even the PP chemotype (ggHH), indicating that neither the dominant gene G nor H is necessary for the enzymatic formation of GPP. On the other hand, the limonene synthase activity was detected only in the plants of the PA chemotype (GGHH) which are capable of producing the cyclic monoterpenoid, perillaldehyde. These results indicated that the coexistence of the two genes, G and H, is absolutely required for separate enzymatic reactions in the cyclization of GPP to yield l-limonene, an intermediate in the biosynthesis of perillaldehyde. It has been suggested that G may be responsible for the ionization of GPP, while H may be required for the coupled reaction steps of isomerization and cyclization which should lead to the formation of l-limonene.

The genetic studies mentioned above have revealed interesting facts that the initiation of the entire biosynthesis of cyclic as well as acyclic monoterpenoids in Perilla plants is collectively controlled by a single dominant gene, G, and that the diversity of chemical compositions of essential oils among the Perilla varieties occurs because of differences in the genotypes for several genes that control various reaction steps in the biosynthetic pathways.

Variation in Leaf Color

The cultivated plants of Perilla have been classified into the so-called purple leaf variety (*Perilla frutescens* Britt. var. *acuta* Kudo) and the green leaf variety (*Perilla frutescens* Britt. var. *acuta* f.*viridis* Makino), according to the ability to accumulate anthocyanins (cyanin and its *p*-coumaric acid ester) in the epidermis on both sides of the leaf. In addition to these forms, there is another form (*Perilla frutescens* Britt. var. *acuta* Kudo f. *crispidiscolor* Makino) in which only the reverse side of the leaf is colored purple in contrast to the green obverse side.

The purple leaves are used in Japan as a dye for coloring pickled plum, aubergine and cucumber, while the green leaves are used as a relish or a garnish, although both the purple and the green leaves usually contain similar essential oils of the perillaldehyde type. For some unknown reason, only the purple leaves have been used as the crude drug for various prescriptions since it was specified in the Chinese Herbal (Li, 1596). It remains an unsettled question whether or not the anthocyanins might have any pharmacological effect, since they are generally regarded as physiologically inactive substances.

In order to study the genetic differences in pigment formation between the purple leaf and the green leaf varieties, Koezuka *et al.* (1984) intercrossed them to obtain the F_1 hybrids. In their F_2 progenies, the purple leaf vs. the green leaf plants segregated in a 3:1 ratio, indicating that a dominant gene, A, is responsible for the anthocyanin formation in the leaves (Koezuka *et al.*, 1985a). It was also demonstrated that the purple pigmentation occurs on both sides of the leaf in the plants having another dominant gene, K, in the presence of A, whereas it takes place exclusively in the reverse side of the leaf in the plants with the genotype AAkk in the absence of the dominant allele K.

These genetic analyses have shown that the anthocyanin formation in Perilla is controlled by a 'master' gene, A, while the site of its expression in the leaf is determined by another dominant gene, K. Thus, it appears that the difference between the purple and the green varieties of Perilla must be very small as far as the anthocyanins formation is concerned. If there were any distinct difference in biological activity between the two varieties, as the herbalists had claimed, it might have been due to either the anthocyanin pigment controlled by the gene A or an unknown substance controlled by a gene that might be very closely linked to the dominant allele A instead of the recessive allele a. Furthermore, Saito (1995) has recently demonstrated by the analysis of random amplified polymorphic DNA (RAPD) of the genomic DNA that the difference in genetic background between the purple leaf variety (var. *crispa* f. *crispa*) and the green leaf variety (var. *crispa* f. *viridis*) was very small despite the difference in the production pattern of anthocyanins.

In view of wide diversity in chemical characters such as oil components and pigments in Perilla, it is considered to be necessary to establish a stable cultivated variety with a definite genotype to make the supply of a uniform and safe crude drug of good quality to pharmacologists as well as to pharmaceutical manufacturers possible.

PHARMACOLOGY OF PERILLA

Historical Background

The earliest record of the medicinal use of *Perilla frutescens* is found in "Jingui Yaolue" (Zhang Zhong-jing, *ca* 219), one of the oldest books of Chinese medicine (Kanpo medicine) compiled in the late Hang dynasty, which briefly stated that drinking the decoction of either the leaf or the seed of Perilla would detoxify the poisoning from eating crab. In the Liang dynasty, Tao Hong-jing stated in the herbal "Mingyi Bielu" (the 6th century) that the seed as well as the leaf of Perilla was effective against asthma and palsy. His opinions were adopted by the books of material medica compiled by imperial command in further generations, e.g., "Takwan Pentsao" edited by Tang Chen-wei (1108).

In the Ming dynasty, Li Shi-zhen (1596) described in his encyclopedic materia medica, "Pentsao Kangmu", that Perilla had been claimed by his predecessors as well as by himself to cure such ailments as intermittent fever, poor circulation, abdominal dropsy, acute gastroenteritis, leg cramps, indigestion, constipation, inflammation, phlegm, pain, cough, fish or crab poisoning, and a bite from a snake or a dog. He emphasized that Perilla was a very important drug of modern times.

In present-day China, Perilla leaves (Zisu) are used for driving away a cold, nasal congestion, cough, vomit, and fish or crab poisoning (dose: 4.5–9g), whereas the seeds (Zisuzi) are prescribed in cough, phlegm, asthma, oppressive feeling in one's chest, and constipation (dose: 3–10g), according to "Zhongyao Zhi" (Chinese Materia Medica, 1981, 1988).

Perilla as a Chinese crude drug was probably introduced to Japan by a Japanese envoy of doctors to China during the 7th century. Though it is not clear when the Perilla plant, which is presumed to be native of Himalaya, Burma and China (Hotta *et al.*, 1989), had been introduced to Japan for cultivation, its names in both Chinese, 'So', and ancient Japanese, 'Inue, Norae', appeared in the oldest Japanese herbal "Honzo Wamyo" (*ca* 918) (Aoba, 1991). It is also found, by the name of Shiso (purple Perilla), among the Chinese drugs listed in the "Engishiki" (927), which established laws concerning ceremonies, systems, etc. to be practiced by the imperial court.

Perilla is usually prescribed together with other crude drugs when used in the Kanpo medicine, which is the classic Chinese medicine adopted by the Japanese doctors. For example, a powdered medicine called Xiangsu San (Kohsosan in Japan), originally described in the Chinese Pharmacopoeia in 1110 for curing a cold in its early stages, is composed of five crude drugs: Perillae Herba, Cyperi Rhizoma, Glycyrrhizae Radix, Zingiberis Rhizoma, and Aurantii Nobilis Pericarpium. Interestingly, a leading pharmaceutical industry in Japan added Kohsosan to a Western medicine to fortify the latter's effect on a cold. On the other hand, the leaves and seeds of Perilla have been used singly for folk remedies in Japan; the decoctions are drunk for colds, poisoning from eating fish or crab, acute enteritis, diarrhea, cough, cerebral anemia, etc., while the crumpled fresh leaves are applied on bleeding wounds, ringworm, and athlete's foot. Furthermore, the fresh herb is used locally as a bath medicine for poor blood circulation, rheumatism, neuralgia, and lumbago (Izawa, 1980).

However, Kagawa(1729), a celebrated doctor of Kanpo medicine in the Edo period, doubted of such multifold clinical effects of Perilla as had been mentioned by Li (1596), and emphasized that Perilla could only relieve patients from a cold to some degree. As regards the kind of Perilla to be used as a remedy, both Chinese herbalists of old times and Japanese doctors, such as Naitoh (1842), insisted that only a variety with purple crinkled leaves emitting aroma should be selected, rejecting the green or non-aromatic leaves. Despite the historical records about Perilla found in the age-old Chinese herbals as well as in the Japanese folklore, only a few pharmacological studies had been made until recently, since Perilla was commonly regarded a spice rather than a drug , showing no drastic effects but stimulating one's appetite. In recent years, however, efforts have been made to verify some of the traditions concerning Perilla scientifically by undertaking the bioassays of the leaf and seed extracts. Significant information obtained from these studies are reviewed here according to their biological activities.

Antimicrobial Activity

The antibacterial activity of Perilla leaves, as might have been expected from their uses for detoxifying the fish and crab poisons in China and for preserving pickles and garnishing a dish of sashimi (fresh slices of raw fish) in Japan, was reported by Okazaki *et al.*(1951) who showed that the water extract of the leaves was active against *Staphylococcus aureus*. Honda *et al.* (1984), however, found that the Perilla leaf extract was inactive against Gram-negative bacteria, such *as Escherichia, Pseudomonas, Citrobacter,* and *Serratia*, but its ethereal fraction, not aqueous fraction, showed a weak growth-inhibiting activity against such Gram-positive bacteria as *Sarcina* and *Bacillus* at a concentration as high as 1.6mg/ml. Unexpectedly, perillaldehyde (0.4mg/ml), the main component of the essential oil, hardly inhibited the growth of all the nine strains of bacteria including *Staphylococcus aureus*, a result contradictory to an earlier report by Xu (1947).

Interestingly, however, the ethereal fraction of the Perilla leaf extract strongly inhibited the growth of six species of such dermatophytic fungi as *Trichophyton, Microsporum, Sabourandites,* and *Epidermophyton,* though it was ineffective against *Saccharomyces* and *Candida* (Honda *et al.*,1984). It was shown that the antidermatophytic activity was partly due to perillaldehyde which was inhibitory to the growth of those fungi at a concentration higher than 0.1 mg/ml. These results seem to be concordant with the fact that the fresh leaves of Perilla used to be applied on the *Trychophyton*-infected athlete's foot and ringworm in folk medicine, even if its medical effectiveness in practice might be questionable.

Kang *et al.* (1992) have demonstrated that a steam distillate of fresh leaves of *Perilla frutescens* cultivated in California showed a moderate growth-inhibiting activity against 14 of 16 species of bacteria and fungi tested, at the minimum inhibition concentrations (MIC) ranging from 31.2 to 1000 µg/ml. Interestingly, the steam distillate inhibited the growth of the Gram-negative *Salmonella choleraesuis* (MIC: 500 µg/ml), which is one of the major bacteria that cause food poisoning. The steam distillate mainly consisted of perillaldehyde (74.0%), limonene (12.8%), β-caryophyllene (3.8%), α-bergamotene (3.5%), and linalool (2.6%). Perillaldehyde among others showed a wide spectrum of antimicrobial activity (MIC: 125–1000 µg/ml) similar to that of the steam distillate. Furthermore, a combination of perillaldehyde and polygodial, a sesquiterpenoid from the leaves of *Polygonum hydropiper*, showed synergistic effects against both bacteria and fungi. It was suggested that the customary addition of the leaves of *Perilla frutescens* and *Polygonum hydropiper* to raw fish slices (sashimi) may be beneficial in the prevention of *Salmonella* poisoning.

Effects on the Central Nervous System

To provide a pharmacological basis for prescribing dried leaves of Perilla for neurosis in Kanpo medicine, Sugaya *et al.* (1981) studied the depressive effects of an aqueous extract (AE) and perillaldehyde (PA) on the central nervous systems of various animals. The positive results obtained were as follows: (1) a decrease in motility was observed in the rats to which AE (4 g/kg) was orally administered, although PA was ineffective; (2) the nervous reflex in the upper larynx of cats was inhibited by an intravenous injection of either AE (400 mg/kg body weight) or PA (50 mg/kg); (3) the hexabarbital-induced sleep in mice was significantly prolonged by the oral administration of AE (4 g/kg) as well as of PA (100 mg/kg).

Such a prolongation of hexabarbital-induced sleep was observed in mice administered orally with a methanol extract (2 g/kg) of dried leaves of Perilla (Honda *et al.*, 1986). However, its effect widely varied with the chemotype of Perilla from which the extract was prepared; the sleeping time was prolonged by 52% for the L-PA type, 80–90% for the PA, PK and EK types, 170% for the PP-M type, and 380% for the PP-DM type. No case of death was observed in the mice treated with these extracts, except for the extract prepared from the leaves of the PK chemotype that caused death in all the treated mice within 24 hr, the toxicology of which will be mentioned later.

Attempts were made to isolate the active principles responsible for the sleep prolongation from the leaves of the PA type through the bioassays of many fractions of the extract on mice (Honda *et al.*, 1986). The results indicated that the sleep was prolonged in the presence of two structurally unrelated compounds, perillaldehyde and stigmasterol. Interestingly, the sleep-prolonging activity was not observed when perillaldehyde (2.5 mg/kg, *per os*) or stigmasterol (0.75 or 5 mg/kg, *per os*) was singly given to the mice, but only when they were simultaneously administered to the mice, which increased the sleeping time by 43%. Similarly, a decrease in the motility and body temperature was observed only in the mice treated simultaneously with the two compounds. These findings suggest that the sedative activity of the Perilla leaf is due to an interaction between perillaldehyde, a monoterpenoid specific to the plant of the PA type, and sigmasterol, a phytosterol widely distributed among higher plants. It has been suggested that stigmasterol, which is known to increase the fluidity of cell membranes, might potentiate the action of perillaldehyde on the central nervous system. In this connection, it is reasonable that the doctors used an aromatic variety of *Perilla*, which must have contained perillaldehyde.

On the other hand, another sleep-prolonging substance isolated from the plants of the PP chemotype producing no monoterpenoids was identified as dillapiole, a phenylpropanoid characteristic of this chemotype (Honda *et al.*, 1988). It showed a significant activity on mice at a small dose (1.13 mg/kg *per os*), being four times as active as myristicin that was present together with dillapiole in the PP-DM type. Since both the compounds are potent psychotropic substances (Shulgin, 1966, 1967), it will be unsafe to use the leaves of the PP type as well as those of the PK type for spice and remedy.

Promotion of Intestinal Propulsion

Koezuka *et al.* (1985b) investigated effects of the Perilla leaves of various chemotypes on the excretory activity of intestines, on the presumption that it might be related to such antidotal capability of this herb against fish- and crab-poisoning as described in the old herbals. As a result, only the leaf extracts from plants of the perillaketone (PK) type were found to promote intestinal propulsion, which was measured by the movement of barium sulfate in the small intestine of mice. Furthermore, the active principle isolated was determined to be perillaketone, a major monoterpenoid of the PK chemotype. Perillaketone increased intestinal propulsion by 30% at a dose of 15 mg/kg body wt., *per os*, and shortened the time required for the excretion of a charcoal meal fed to mice as effective as 60 mg/kg castor oil or 100 mg/kg magnesium sulfate, which is used as a purgative for food poisoning.

Pharmacological studies on the mechanism of this compound suggested that the promotion of intestinal propulsion, which is accompanied with an increasing rate of the transport of contents through the intestinal canal, is largely due to stimulation of the motility of circular muscles of the intestine. Since the plants of the PA chemotype do not contain perillaketone but perillaldehyde, there is a possibility that some other constituents of these leaves and seeds of Perilla might be responsible for the legendary, antidotal effect on food poisoning.

Toxicity of Perillaketone

Wilson *et al.* (1977) reported that perillaketone caused serious lung edema in the cattle grazing wild Perilla plants in the pasture. The toxicity of this compound has also been indicated by the experiments using mice (Koezuka *et al.*, 1985b). These findings seem to support the validity of the traditional practice of using a *Perilla* variety of the PA type instead of the PK type for medical purpose. Fortunately, it is required by the Japanese Pharmacopoeia that the Perilla leaf of the PA type should be adopted as a remedy for Kanpo Medicine. Actually, Nakagawa *et al.* (1993) have reported that the mice administered with large doses (450 and 900 mg/kg, *per os*) of perillaldehyde in corn oil showed no abnormalities in the weight and morphology of body, liver, and kidney as well as in the activities of serum GOT and GPT, the marker enzymes for a liver trouble.

In connection with the alarming report (Wilson *et al.*, 1977) of the death of many cattle that grazed on Perilla leaves containing perillaketone (PK), a pulmonary emphysema-inducing agent, Takano *et al.* (1990) have investigated the distribution of PK in Perillae Herba put on the market as a crude drug used for Kanpo medicine. For the detection of PK in the dried leaves, the ethyl acetate extracts were analyzed by capillary GC-MS under optical conditions, which enabled them to determine the quantity of PK as small as 0.001 ng. The results showed that PK was detected in 9 of the 25 samples (36%) of Japanese, Chinese, and Korean products, most frequently in the imports from China. The content of PK varied widely from 1.1 to 119.9 mg/100g of dry leaves. The authors argued that the final content of PK in a decoction used for Kanpo medicine would be negligible, since the common dose of Perillae Herba to be taken by patients is 1 to 3 g/day and only about 3% of the essential oils remain in the preparation after boiling. Nevertheless, the authors do not deny the possible danger of inhaling PK evaporating from the decoction during the manufacturing process. On the other hand, in the form of powdered drug which contains 0.4–0.8 g of Perillae Herba as a daily dose, the maximum intake of PK would amount to 0.96 mg in case the sample with the highest content of PK is used, although this value is one thousandth of the LD_{50} value reported for the mice by Wilson *et al.* (1977). The toxic dose of PK to humans, however, is still unknown.

Antitumor Activity

Samaru *et al.* (1993a, b) studied the influence of 36 kinds of spices including Perilla leaf on the survival of mice inoculated with the MM2 ascites tumor. In the experiments, feed containing a relatively small amount (0.1–1.0%) of spice was supplied freely to each group of ten mice. As a result, one or two of the ten mice fed with any one of Perilla, wasabi (Japanese horseradish), basil, sesame, sage, fennel, and peppermint survived for more than 39 days, while the rest of the treated mice as well as the control animals without spice could survive only for about 12 days on the average after the inoculation of tumor (two million cells/mouse) into the abdominal cavity. Surprisingly, two of the ten mice on a feed mixed with 1% Perilla leaf did survive for more than 280 days.

Although the active principles of these spices are still unknown, Samaru *et al.* (1993a) have speculated about the sporadic survivals of the mice that certain spices would bring about such good results by affecting tumor cells, not directly, but indirectly, making some alterations to the physiological condition, inner environment, and metabolism of tumor cells, or to the "allelopathy" between the tumor and the host through unknown mechanisms.

Perilla seed oil, which is rich in such n-3-polyenoic fatty acids as eicosapentaenoic acid and docosahexaenoic acid, is reportedly inhibitory to carcinogenicity in the large intestine of rats (Narusawa *et al.*, 1990). Hori *et al.* (1987) reported a remarkable effect of Perilla seed oil on the pulmonary metastasis of ascites tumor cells in rats. They had kept feeding three-week-old rats with a diet supplemented with either 5% Perilla oil or 5% safflower oil, before they injected Yoshida sarcoma cells (10^4 cells/rat) into the abdominal cavity of five-week-old rats. Two weeks later, they found that the number of metastatic foci observed on the lung surface was significantly smaller in the rats fed with Perilla seed oil (average: 3.8 foci/rat) than in the rats fed with safflower oil (7.0 foci) or with a normal diet without any vegetable oil (8.8 foci). It was suggested that the fatty acids of Perilla seed oil, which is very rich in α-linolenic acid (64.0% of the seed oil), modify not only the host animal cells but also the metastatic potentials of the ascites tumor cells.

Inhibition of TNF-α Overproduction

Tumor necrosis factor-α (TNF-α), a protein secreted by macrophages, not only shows a strong necrotic activity on tumor cells but also plays an important role in the process of inflammation. However, the continuous overproduction of TNF-α is known to cause damage to tissues, aggravating inflammation. In an attempt to screen for vegetables capable of inhibiting the overproduction of TNF-α, Ueda and Yamazaki (1993) found such an activity in the leaf extract of *Perilla frutescens* var. *acuta* Kudo. In this study, the dried leaves of Perilla were homogenized in water and the filtrate (400μl) was administered orally to mice that had been treated with both priming and triggering agents to induce a high level of TNF in the blood. As a result, the level of TNF in the treated mice was decreased by 86%. But the chemical principle for the TNF-inhibiting activity has not been clarified, although it was supposed to be a heat-stable substance. The inhibitory activity of the Perilla extract on TNF overproduction was as high as that of the decoction of "Saibokuto", a traditional prescription, which consists of ten kinds of crude drugs including Perilla leaf and is used for such ailments as asthma, bronchitis, cough, and neurosis. Incidentally, according to a report of clinical tests carried out in China, a 25% decoction prepared from a mixture of Perilla leaf and dried ginger (10:1), when given to 552 patients suffering from chronic bronchitis at a daily dose of 100 ml in a 10-day cycle, was more or less effective on 363 (65.8%) of the patients (Jiangsu New Medical College, ed., 1977).

Influence of Perilla Seed Oil on Lipid Metabolism

Perilla seeds (fruits), which have been used mainly for getting rid of cough and phlegm in traditional Chinese medicine, are also said to be useful for the prevention of arteriosclerosis by Japanese folklore, if a tea prepared from a pinch of seeds is taken daily (Iwao and Kobayashi, 1980). The seed oil of Perilla has lately attracted considerable attention as a health food, because it is rich in unsaturated fatty acids, in particular, linolenic acid which is essential for maintaining health. Yamamoto *et al.* (1987, 1988) have demonstrated that the rats that had been fed a diet supplemented with Perilla oil rich in α-linolenic acid through two generations showed an increase in the proportion of docosahexaenoic acid in the brain phospholipids, as compared with the safflower oil-fed rats. Furthermore, in the brightness-discrimination learning ability test, the Perilla oil groups were found to be superior in the correct response ratio than the safflower oil groups.

Nanjo *et al.* (1993) have shown that the proportions of the 18:3, 18:2, and 18:1 unsaturated fatty acids in Perilla oil were 52.6, 15.5, and 17.7%, respectively, in marked contrast to 0, 9.8, and 37.5% in palm oil. Consequently, the levels of cholesterol, phospholipids, and triglycerols in the plasma of rats proved to become remarkably low when they were kept on a 30% Perilla oil diet over a period of 31 days instead of a 30% palm oil diet.

Biochemical studies on the dietary manipulation by Perilla seed oil of hepatic lipids and its influence on peroxisomal β-oxidation and serum lipids have been carried out by Kawashima and Kozuka (1993a, b). Rats and mice were fed with a medium fat diet (11.6% Perilla oil) or a high fat diet (18.8% Perilla oil) for four weeks before they were sacrificed for analysis. The results showed that Perilla oil feeding enhanced not only the content of α-linolenic acid (18:3) but also the biosynthesis of eicosapentaenoic acid (20:5) and docosapentaenoic acid (22:6) in the liver of both animals. Perilla oil stimulated the specific activity of peroxysomal β-oxidation as much as clofibric acid (50 mg/kg body wt.) did which is a hypolipidemic drug known to stimulate the proliferation of peroxisomes. However, Perilla oil was less effective than fish oil in reducing the levels of serum cholesterol and circulating triacylglycerol, despite a quite high proportion of docosahexaenoic acid in the hepatic lipid.

Kawashima *et al.* (1994) also have demonstrated that feeding diets supplemented with either Perilla oil or fish oil to rats over a two-week period caused a marked decrease in the content of arachidonic acid (20:4), a precursor in prostaglandin biosynthesis, in *hepatic* phosphatidylcholine and phosphatidylethanolamine. Feeding rats with Perilla oil and fish oil also reduced the content of arachidonic acid in *serum* phospholipid by 67 and 77%, respectively. The level of arachidonic acid in *renal* phosphatidylcholine appeared to be regulated by that in *hepatic* phosphatidylcholine through changes in the *serum* level of arachidonic acid. It was suggested that Perilla oil might indirectly modulate prostaglandin formation in kidney by altering the acyl composition of *hepatic* phospholipids. Actually, the formation of prostaglandin E_2, the most biologically potent of mammalian prostaglandins, decreased by 75% in the kidney of Perilla oil-fed rats. The active principles responsible for these effects need to be determined by analysing the fractions of Perilla seed oil.

Antiallergic Activity

The antiallergic and anti-inflammatory activities of the leaf extract of *Perilla frutescens* have been reviewed by Yamazaki (1993), while its useful application to food industry has been considered by Kosuna (1994). The details of these studies are described in another chapter of this book.

Allergic Contact Dermatitis

Contact dermatitis due to *Perilla frutescens* is well known in Japan as an occupational dermatitis. Kanzaki and Kimura (1992) reported three cases of dermatitis in women who had made contact with Perilla leaves for a few years while wrapping up the fruits of *Prunus mume* in the leaves to make fragrant and red-colored pickles. These women developed such allergic symptoms as vesicular eruption, diffuse erythema, mild edema, and marked hyperkeratosis with fissures on their fingers, especially on the finger tips, which showed positive reactions to the juice extracted from Perilla leaves in patch tests. The hand dermatitis in one of the three women was cleared on treatment with a topical corticosteroid cream in 10 days. It is considered that the major chemical constituent responsible for the dermatitis is perillaldehyde which apparently has a strong sensitizing potential. In the production of a Perilla extract which has been shown to be effective in preventing and treating allergic diseases, perillaldehyde was removed from the final products (see also Chapter 8, this book).

CONCLUSION

The biological activities of Perilla leaves and seeds, which have been used as crude drugs for thousands of years in the East, are quite complex and manifold, as is often the case with traditional Chinese remedies. Furthermore, the pharmacological effects of Perilla leaves may vary with differences in varieties, in particular, the chemotypes for essential oils. Although chief attention has been paid on the pharmacological activities of the essential oils of leaves, it is also being directed to those of the aqueous extract and the seed oil. Recent studies on pharmacological and biochemical mechanisms of these materials have deepened our understanding of this interesting herb. In many cases, however, the active principles have not yet been clarified, but hopefully they will be elucidated by chemical analyses combined with appropriate bioassays for the individual fractions of leaf and seed extracts. It should be kept in mind that a particular activity is not necessary due to a single constituent of the crude extract but might be expressed only in the presence of two or more compounds, as was observed for the sedative activity of Perilla. The occurrence of various types of interactions between different constituents is considered as a characteristic of Chinese crude drugs, which may be expected to give well-balanced, milder pharmacological actions with less side effects compared with single natural compounds or synthetic drugs. Nevertheless, further efforts must be made to isolate and identify the potential, bioactive compounds from Perilla leaves and seeds in order to fully understand the pharmacological activities and their mechanisms at the molecular levels.

REFERENCES

Aoba, T. (1991) *Yasaino Nihonshi (History of Vegetables in Japan)*. Yasaka Shobo, Tokyo, 317 pp. (in Japanese).

Chinese Academy of Medical Science (eds.) *Zhongyao Zhi (Chinese Materia Medica)*. Vol. **3**, 1981, 781 pp.; Vol.**4**, 1988, 862 pp, People's Hygiene Publishing Co., Beijing.

Honda, G., Koga, K., Koezuka, Y. and Tabata, M. (1984) Antidermatophytic compounds of *Perilla frutescens* Britton var. *crispa* DECNE. *Shoyakugaku Zasshi*, **38**, 127–30 (in Japanese).

Honda, G., Koezuka, Y., Kamisako, W. and Tabata, M. (1986) Isolation of sedative principles from *Perilla frutescens. Chemical and Pharmaceutical Bulletin*, **34**, 1672–1677.

Honda, G. Koezuka, Y. and Tabata, M. (1988) Isolation of dillapiole from a chemotype of *Perilla frutescens* as an active principle for prolonging hexobarbital-induced sleep. *Chemical and Pharmaceutical Bulletin,* **36**, 3153–3155.

Hori, T., Moriuchi, A., Okuyama, H., Sobajima, T., Tamiya-Koizumi, K., and Kojima, K. (1987) Effect of dietary essential fatty acids on pulmonary metastasis of ascites tumor cells in rats. *Chemical and Pharmaceutical Bulletin,* **35**, 3925–3927.

Hotta, M., Ogata, K., Nitta, A., Hosikawa, K., Yanagi, M. and Yamazaki, K. (eds.) (1989) *Useful Plants of the World*. Heibonsha, Tokyo, 1499 pp. (in Japanese).

Ito, H. (1970) Studies on Folium perillae. VI. Constituent of essential oils and evaluation of Genus *Perilla. Yakugaku Zasshi*, **90**, 883–892 (in Japanese).

Iwao, H. and Kobayashi, M. (1980) *Yasai wa Kusuri da (Vegetables are Drugs)*. Nosan Gyoson Bunka Kyokai, Tokyo, 254 pp. (in Japanese).

Izawa, B. (1980) *Illustrated Cyclopedia of Medicinal Plants of Japan*. Seibundo Shinkosha, Tokyo, 331 pp. (in Japanese).

Jiangsu New Medical College (ed.) (1977) *Dictionary of Chinese Materia Medica*. Vol. **2,** Shanghai Science-Technology Publishing Co., Shanghai, 2754 pp. (in Chinese).

Kagawa, S. (1729) *Ippondo Yakusen,* Edited by Nanba, T., Kanpo Bunken Kankokai, Osaka, 1976, 858 pp. (in Chinese).

Kang, R., Helms, R., Stout, M.J., Jaber, H., Chen, Z. and Nakatsu, T. (1992) Antimicrobial activity of the volatile constituents of *Perilla frutescens* and its synergistic effects with polygodial. *J. Agric. Food Chem.,* **40**, 2328–2330.

Kanzaki, T. and Kimura, S. (1992) Occupational allergic contact dermatitis from *Perilla frutescens* (shiso). *Contact Dermatitis*, **26**, 55.

Kawashima, Y., Musoh, K. and Kozuka, H. (1993a) Alterations by clofibric acid of glycerolipid metabolism in rat-kidney. *Biochimica et Biophysica Acta,* **1169**, 202–209.

Kawashima, Y. and Kozuka, H. (1993b) Dietary manipulation by Perilla oil and fish oil of hepatic lipids and its influence on peroxisomal β-oxidation and serum lipids in rat and mouse. *Biological and Pharmaceutical Bulletin,* **16**, 1194–1199.

Kawashima, Y., Mizuguchi, H. and Kozuka, H. (1994) Modulation by dietary oils and clofibric acid of arachidonic acid content in phosphatidylcholine in liver and kidney of rats: Effects on prostaglandin formation in kidney. *Biochimica et Biophysica Acta,* **1210**, 187-194.

Koezuka, Y., Honda, G. and Tabata, M. (1984) Essential oil types of the local varieties and their F$_1$ hybrids of *Perilla frutescens. Shoyakugaku Zasshi,* **38**, 238–242 (in Japanese).

Koezuka, Y., Honda, G., Sakamoto, S. and Tabata, M. (1985a) Genetic control of anthocyanin production in *Perilla frutescens. Shoyakugaku Zasshi,* **39**, 228–231 (in Japanese).

Koezuka, Y., Honda, G., and Tabata, M. (1985b) An intestinal propulsion promoting substance from *Perilla frutescens* and its mechanism of action. *Planta Medica,* 1985, 480–482.

Koezuka, Y., Honda, G., and Tabata, M. (1986a) Genetic control of the chemical composition of volatile oils in *Perilla frutescens. Phytochemistry,* **25**, 859–863.

Koezuka, Y., Honda, G., and Tabata, M. (1986b) Genetic control of phenylpropanoids in *Perilla frutescens. Phytochemistry,* **25**, 2085–2087.

Koezuka, Y., Honda, G., and Tabata, M. (1986c) Genetic control of isoegomaketone formation in *Perilla frutescens. Phytochemistry,* **25**, 2656–2657.

Kosuna, K. (1994) Regulation of biodefence mechanism by food. *New Food Industry,* **36,** 41–44 (in Japanese).

Li, Shi-zhen (1596) *Pentsao Kanmu (Honzo Kohmoku).* Edited by Kimura, K. and translated into Japanese by Suzuki, S.), Vol.**4,** Shunyodo, Tokyo, 1979, 650 pp.

Naito, S. (1842) *Koho Yakuhin Koh.* A reprinted edition, Ryogen, Tokyo, 1974, 465 pp.

Nakagawa, Y., Tayama, K., Miyakawa, H. and Ichikawa, H. (1993) Effects of perillaldehyde on liver and kidney of mice. *Annual Report of Tokyo Metropolitan Research Laboratory,* **44**, 261–263 (in Japanese).

Nanjo, F., Honda, M., Okushio, K., Matsumoto, N., Ishigaki, F.,Ishigami, T. and Hara, Y. (1993) Effects of dietary tea catechins on α-tocopherol levels, lipid peroxidation, and erythrocyte deformability in rats fed on high palm oil and Perilla oil diets. *Biological and Pharmaceutical Bulletin,* **16**, 1156–1159.

Narusawa, T., Takahashi, M., Kusaka, H., Yamazaki, Y., Koyama, H., Kotana, M., Nishizawa, Y., Koban, M., Isoda, Y. and Hirano, J. (1990) Inhibition of carcinogenesis in the large intestine of rats by perilla oil, a cooking oil rich with a ω-3-polyunsaturated fatty acids, α-linolenic acid. *Igakuno Ayumi,* **153**, 103–104 (in Japanese).

Nishizawa, A., Honda, G. and Tabata, M. (1989) Determination of final steps in biosynthesis of essential oil components in *Perilla frutescens. Planta Medica,* **55**, 251–253.

Nishizawa, A., Honda, G., and Tabata, M. (1990) Genetic control of perillene accumulation in *Perilla frutescens. Phytochemistry,* **29**, 2873–2875.

Nishizawa, A., Honda, G., and Tabata, M. (1991) Genetic control of elsholtziaketone formation in *Perilla frutescens. Biochemical Genetics,* **29**, 43–47.

Nishizawa, A., Honda, G. and Tabata, M. (1992a) Genetic control of peltate glandular trichome formation in *Perilla frutescens. Planta Medica,* **58**, 188–191.

Nishizawa, A., Honda, G. and Tabata, M. (1992b) Genetic control of the enzymatic formation of cyclic monoterpenoids in *Perilla frutescens. Phytochemistry,* **31**, 139–142.

Okazaki, K. and Wakatabe, T. (1951) Antibacterial activity of higher plants. XIV. Antibacterial activity of crude drugs (4). *Yakugaku Zasshi,* **71**, 481–482 (in Japanese).

Saito, K. (1995) Molecular engineering of plant metabolism. *Abstract of 32nd Symposium on Phytochemistry,* Tohoku University , Sendai, p. 48–58 (in Japanese).

Samaru, Y., Hanada, S. and Sudo, K. (1993a) Allelopathy between the cancer and the host. II-A. Influence of spices on the ascites tumor in mice. *Minophagen Medical Review,* 1993 MAR., 48–55 (in Japanese).

Samaru, Y., Hanada, S. and Sudo, K. (1993b) Allelopathy between the cancer and the host. II-B. Influence of spices on the ascites tumor in mice. *Minophagen Medical Review,* 1993 MAY, 45–51 (in Japanese).

Shulgin, A.T. (1966) Possible implication of myristicin as a psychotropic substance. *Nature,* **210,** 380–384.

Shulgin, A.T. (1967) Psychotropic phenylisopropylamines derived from apiole and dillapiole. *Nature,* **215,** 1494–1495.

Sugaya, A., Tsuda, T. and Obuchi, T. (1981) Pharmacological studies on Perillae Herba I. Neuropharmacological action of water extract and perillaldehyde. *Yakugaku Zasshi,* **101,** 642–648. (in Japanese).

Takano, T., Yasuda, I., Hamano, T., Seto, T. and Akiyama, K. (1990) Determination of perilla ketone in Perilla herb by capillary gas chromatography/mass spectrometry. *Eisei Kagaku,* **36,** 320–325.

Tang, Chen-wei (1108) *Chingsu Chenglei Takwan Pentsao* (edited by Ai, S.). Reproduced from the original text by Hirokawa Publishing Co., Tokyo, 1970, 750 pp. (in Chinese).

Tao, Hong-jing (the 6th century) *Mingyi Bielu* (edited by Shang, Z.). Reproduced by People's Hygiene Publication Co., Beijing, 1986, 342 pp. (in Chinese).

Ueda, H. and Yamazaki, M. (1993) Inhibitory activity of perilla juice for TNF-α production. *Japanese J. of Inflammation,* **13,** 337–340 (in Japanese).

Wilson, B.J., Garst, J.E., Linnabary, R.D., and Channell, R.B. (1977) Perilla ketone: a potent lung toxin from the mint plant, *Perilla frutescens* Britton. *Science,* **197,** 573–574.

Xu, Z. (1947) *Nong Bao* **1,** 17 (cited from *Zhongyao Zhi,* **4,** 671, People's Hygiene Publishing Co., Beijing, 1988).

Yamamoto, N., Saitoh, M., Moriuchi, A., Nomura, M. and Okuyama, H (1987) Effect of the dietary α-linolenate/ linoleate balance on brain lipid compositions and learning ability of rats. *J. of Lipid Research,* **28,** 144–151.

Yamamoto, N., Hashimoto, A., Takemoto, Y., Okuyama, H., Nomura, M., Kitajima, R., Togashi, T. and Tamai, Y. (1988) Effect of the dietary α-linolenate/linoleate balance on lipid compositions and learning ability of rats. II. Discrimination process, extinction process, and glycolipid compositions. *J. of Lipid Research,* **29,** 1013–1021.

Yamazaki, M. (1993) Anti-inflammatory and antiallergic activities of *Perilla* juice. *Fragrance Journal,* **No. 9,** 75–81 (in Japanese).

Yuba, A., Honda, G., Koezuka, Y. and Tabata, M. (1995) Genetic Analysis of essential oil variants in *Perilla frutescens. Biochemical Genetics,* 33, 341–348.

Zhang, Z. (*ca* 219) *Kinki Yoryaku* (edited by Research Institute for Chinese Medicine; translated by Suzuki, T. into Japanese from the original text, *Jinkui Yaolue*), Chugoku Kanpo, Tokyo, 1982, 476pp.

12. MOLECULAR BIOLOGY IN *PERILLA FRUTESCENS*
(Isolation of Specifically Expressed Genes in Chemotypes)

MAMI YAMAZAKI, ZHI-ZHONG GONG and KAZUKI SAITO

*Faculty of Pharmaceutical Sciences, Laboratory of Molecular Biology
and Biotechnology in Research Centre of Medicinal Resources,
Chiba University, Yayoi-cho 1-33, Inage-ku, Chiba 263, Japan*

INTRODUCTION

Perilla frutescens is an annual herb used as a traditional Chinese medicine, and for garnish and food colouring. This plant has been widely cultivated in Japan. Several cultivars of *P. frutescens* are known such as "Akashiso" (purple leaf), "Aoshiso" (green leaf), "Chirimenshiso (crispy purple leaf)", "Chirimen-aoshiso (crispy green leaf). Additionally, several chemotypes of different composition of essential oils are also known. These phenotypes related to the composition of secondary metabolites are genetically stable, and each chemotype can be crossed to other chemotypes. Thus, genetic analyses of inbred lines of these chemotypes have been done successfully (Koezuka *et al.*, 1985, 1986; Yuba *et al.*, 1995). Recently, the techniques of molecular biology have been used to investigate the biosynthetic pathway of secondary products in this species. The genes involved in the biosyntheses of anthocyanin (Yamazaki *et al.*, 1996) and monoterpene (Yuba *et al.*, submitted) were cloned from these chemotypes.

GENETIC STUDIES OF *P. FRUTESCENS*

In the purple leaf cultivar, "Akashiso", anthocyanin is accumulated in the epidermis of the leaf and stem. The main component of anthocyanin in this plant is shisonin (cyanidin 3,5-glucoside). Additionally, there are two other varieties, *e.g.*, "Aoshiso" that contains only a trace amount of anthocyanin and "Katamenshiso" that contains anthocyanin only in the lower epidermis not in upper epidermis. Genetic analyses of these varieties by intercrosses showed that three independent loci (*A, B* and *K*) control the formation of anthocyanins in different parts of the plants (Koezuka *et al.*, 1985). The locus *A* promotes the pigmentation in the leaf and stem. The loci *B* and *K* are responsible for the pigmentation in the epidermis of stem and the upper epidermis of leaf, respectively.

Koezuka *et al.* (1985) have established four inbred lines of chemotype in volatile oil compositions, perillaldehyde- (PA-), elsholtziaketone- (EK-), perillaketone- (PK-), and phenylpropanoid- (PP-) types, by self pollination. The plants of monoterpenoid types (PA-, EK- and PK-types) do not contain phenylpropanoids. Conversely, PP-type plants do not contain any monoterpenoid. The crossing experiments among these types showed that at least two loci (*G* and *H*) control the expression of these four phenotypes. When the genotype is *GG* or *Gg*, monoterpenoids are produced from mevalonic acid to PA, EK, and PK. Only when genotype is *gg*, phenylpropanoids are produced from shikimic acid. The locus *H* promotes the formation of cyclohexene ring in PA. From further genetic studies using additional subtypes, the presence of loci (*D* and *E, Fr1* and *Fr2, I, J, N, P* and *Q, R*) which determine the composition of essential oils have been presumed (Yuba *et al.*, 1995).

ISOLATION OF DIFFERENTIALLY EXPRESSED GENES IN CHEMOTYPES

Differential display of mRNA (Liang and Pardee, 1992) was used to look for the *Perilla* genes involved in anthocyanin biosynthesis. The poly(A)$^+$ RNA populations in the purple leaf variety and the green leaf variety were compared by means of the polymerase chain reaction (PCR) using pairs of an anchored oligo(dT) primer and a short arbitrary primer. The cDNA clones corresponding to the specific bands in the purple leaf variety were isolated and characterised (Yamazaki *et al.*, 1996).

Procedure

Poly(A)$^+$ RNA was isolated from fresh leaves of the dark purple coloured variety 'Shikun' and the green leaf variety 'Seikun'(Sakata, Yokohama, Japan). Then 0.9 µg of poly(A)$^+$ RNA was reverse transcribed in subsets of specific one-base anchored oligo(dT) primers (H-T11G, H-T11A or H-T11C, GenHunter, MA) that recognise different fractions of the total poly(A)$^+$ RNA population in 33 µl of reaction mixture. The resulting cDNA was amplified with the same anchored oligo(dT) primer as used in the reverse transcription and an arbitrary primer (one of H-AP1~8, GenHunter) in PCR with Taq DNA polymerase. The PCR was performed in the 20 µl of reaction mixture which contained of 2 µl cDNA solution, 0.2 µM anchored oligo(dT) primer, 0.2 µM arbitrary primer, 0.12 µM dNTPs, 5 or 10 µCi [^{32}P]dCTP, 10 mM Tris-HCl (pH 9.0), 50 mM KCl, 0.01% TritonX100, 1.25 mM MgCl$_2$ and 1 unit of Taq DNA polymerase. The condition of the PCR was as follows: a 20-sec initial heating at 72°C, followed by 40 three-step cycles of a 30-sec denaturation at 94°C, a 2-min annealing at 40°C, and a 30-sec elongation at 72°C, and then a final 5-min elongation at 72°C in the thermal cycler.

The amplified DNA was fractionated by electrophoresis in sequence gel. The gel was dried and exposed onto X-ray film. The differentially expressed bands were cut out from the gel and eluted with 100 µl H$_2$O by heating at 100°C. The eluted DNA was precipitated with ethanol and resolved in 20 µl H$_2$O. The half aliquot of DNA solution was used as the template for PCR reamplification with the same condition as that which generated the bands. The second PCR condition was the same as in the first PCR amplification except no label was added. Reamplified DNA was separated on 1.5 % agarose gel and the same size of bands as in first PCR were excised and subcloned and used as probes in Northern blot analysis and cDNA screening.

The cDNA library of purple coloured and green varieties of *P. frutescens* was constructed by using λgt10 vector. The 1×10^4 clones of cDNA were screened and 4×10^5 clones were screened for clone A2-3. For Northern blot hybridisation, poly(A)$^+$

Table 1 Specifically amplified fragments in red-colored **(R)** or green **(G)** leaf variety of *P. frutescens*[a] (Yamazaki *et al.*, 1996)

oligo(dT) primers	H-T11G			H-T11A			H-T11C		
	Total	*Specific bands*		Total	*Specific bands*		Total	*Specific bands*	
		R	**G**		**R**	**G**		**R**	**G**
arbitrary primers									
H-AP1	116	0	0	106	0	1 (1)	119	2 (2)	2 (2)
H-AP2	98	0	1 (1)	92	3 (3)	0	135	0	0
H-AP3	102	2 (2)	0	112	4 (4)	0	128	2 (1)	0
H-AP4	117	1 (1)	0	118	7 (7)	1 (0)	84	1 (1)	0
H-AP5	118	3 (3)	0	116	1 (1)	0	106		
H-AP6	117	2 (2)	0	114	0	0	122	2 (1)	0
H-AP7	94	1 (1)	1 (1)	100	0	0	87	0	0
H-AP8	114	3 (3)	1 (1)	93	1 (1)	1 (1)	86	1 (1)	0
Subtotal	876	12 (12)	3 (3)	851	16 (15)	2 (1)	867	8 (6)	3 (3)
Total	Amplified bands: 2,594			Specific bands in R: 36 (33)			Specific bands in G: 9 (7)		

[a] The number of amplified bands were indicated. The number of obtained clones were indicated in brackets.

RNA was denatured and separated in a 1.2% agarose gel containing formaldehyde, followed by transfer to a nylon filter. Hybridisation was performed in 5 × SSPE, 5 × Denhardt's solution, 0.5% SDS, 20 µg/ml denatured salmon sperm DNA at 65°C with ^{32}P-labeled DNA as a probe. The final wash of the filter was performed in 1× SSPE and 0.1% SDS at 65°C for 15 min. To confirm the low level of expression, RT-PCR was performed.

For Southern blot hybridisation, total DNA was extracted from fresh leaves of *P. frutescens* and digested with restriction enzymes (*Eco*RI, *Eco*RV, *Hin*dIII and *Xba*I). Digested DNA (20 µg) was fractionated in electrophoresis on 0.7 % agarose gel and transferred onto nylon filter and then hybridised with ^{32}P-labeled DNA as a probe. The final wash of the filter was performed in 0.1 × SSPE and 0.1% SDS at 65°C for 15 min.

The DNA sequence was determined by double-strand sequencing with the plasmid clones. The result of sequence comparison and molecular analysis was performed using FASTA and BLAST programs at the National Institute of Genetics, Mishima.

Results

The mRNA populations in pigmented and non-pigmented leaf varieties of *P. frutescens* were compared by PCR amplification of poly(A)$^{+}$ using combinations of an anchored oligo(dT) primer and an arbitrary primer. In the amplified band pattern, most bands were present in both varieties. By using 3 oligo(dT) primers and 6 arbitrary primers, 36 fragments were differentially displayed in coloured leaf out of a total of about 2,600 amplified fragments (Table 1). The 33 differential fragments were reamplified from the

Table 2   Isolated cDNA clones from red leaf variety of *P. frutescens*
(Yamazaki *et al.*, 1996)

cDNA clone	Primer combination	Size (kb)	Homologous protein	Expression in **R**	in **G**
A3-5	(H-T11A, H-AP3)	1.4	3GT[a] (maize)	++	-
G8-5	(H-T11G, H-AP8)	1.4	LDOX[b] (grape)	++	-
G8-6	(H-T11G, H-AP8)	1.4	unknown	++	-
A2-3	(H-T11A, H-AP2)	1.0	MYB[c] (Petunia)	+ (RT PCR)	-
G8-33	(H-T11G, H-AP8)	3.2	unknown	nd[d]	nd

[a]3GT – flavonoid-3-*O*-glucosyltransferase
[b]LDOX – leucoanthocyanidin dioxygenase
[c]MYB – homologue of *myb* transcription activator in DNA binding domain
[d]nd – not determined

sequence gel and used as probes in Northern analyses in both varieties. In the Northern hybridisation of poly(A)$^+$ RNA, only 5 of these 33 probes generated specific signals in the purple leaf variety. The cDNA clones (A3-5, G8-5, G8-6, A2-3, G8-33) corresponding to these 5 differentially expressed transcripts were isolated. In the deduced amino acid sequences, cDNA clones A3-5 and G8-5 have 26.3% homology with flavonoid-3-*O*-glucosyltransferase (3GT) of maize and 75.5% homology with leucocyanidin dioxygenase (LDOX) of grape, respectively (Table 2). These deduced proteins are the structural enzymes in anthocyanin biosynthesis, and catalyse the colouring steps of anthocyanins. The deduced protein encoded on the clone A2-3 has 69.9% homology with MYB-homologue of *Petunia* in N-terminus containing DNA-binding domain (Table 2). This clone was expressed in the plant cells at a very low level. In maize, snapdragon and *Petunia*, it has been shown that MYB- and MYC-homologues act as transcriptional activator in anthocyanin biosynthesis (reviewed by Holton and Cornish, 1995). One can assume that the clone A2-3 of *P. frutescens* also promotes the expression of 3GT and LDOX to lead to pigmentation in the leaves. Further investigation is required to determine the biological function of these clones. The clones G8-6 and G8-33 were unknown. In Southern blot analysis, it was shown that two to four copies of each clone presents in the genome of *P. frutescens*. Polymorphism in restricted fragment length between purple and green leaf varieties was detected only in clone A3-5 with *Eco*RV (Data not shown).

CONCLUSIONS

The inbred lines of chemotypes in anthocyanin production and essential oil composition have been established in *P. frutescens*. Genetic analyses of these chemotypes have revealed the presence of genetic loci controlling the biosyntheses of these secondary products (Koezuka *et al.*, 1985, 1986; Yuba *et al.*, 1995). In recent years, the techniques of molecular biology have been used to investigate biosyntheses of secondary metabolites in this species.

Differential display of mRNA is a very sensitive method of detecting low amounts of differentially expressed genes because it is based on PCR. Using this technique for the purple and green leaf varieties, the cDNAs coding 3GT, LDOX and MYB-homologue could be isolated. The 3GT and LDOX would be involved in shisonin biosynthesis. The MYB-homologue might act as a transcriptional activator of these structural genes as well as in maize, snapdragon and *Petunia*. It was shown that differential mRNA display among chemotypes is a powerful tool for the isolation of the genes involved in the production of secondary metabolites in plants. Recently, the cDNA coding limonene cyclase was cloned from PA type of *P. frutescens* by screening using a probe from limonene cyclase of spearmint (Yuba *et al.*, 1996). This gene would be concerned with locus *H*. It is required to connect the isolated cDNAs to the speculated loci in genetic studies.

These results could be applied not only to the breeding of *P. frutescens* but also other medicinal and aromatic plants.

REFERENCES

Holton, T.A. and Cornish, E.C. (1995) Genetics and biochemistry of anthocyanin biosynthesis. *Plant Cell*, **7**, 1071–1083.

Koezuka, Y., Honda, G., Sakamoto, S. and Tabata, M. (1985) Genetic control of anthocyanin production in *Perilla frutescens. Shoyakugaku Zasshi*, **39**, 228–231.

Koezuka, Y., Honda, G. and Tabata, M. (1986) Genetic control of the chemical composition of volatile oils in *Perilla frutescens. Phytochemistry*, **25**, 859–863.

Liang, P. and Pardee, A.B. (1992) Differential display of eukaryotic messenger RNA by means of the polymerase chain reaction. *Science*, **257**, 967–971.

Yamazaki, M., Zhizhong, G. and Saito, K. (1996) Differential display of specifically expressed genes for anthocyanin biosynthesis in *Perilla frutescens. Proc. International Symposium of Breeding Research on Medicinal and Aromatic Plants.* Quedlinburg, June-July 1996.

Yuba, A., Honda, G., Koezuka, Y. and Tabata, M. (1995) Genetic analysis of essential oil variants in *Perilla frutescens. Biochem. Gen.*, **33**, 341–348.

Yuba, A., Yazaki, K., Tabata, M., Honda, G. and Croteau, R. (1996) *Arch. Biochem. Biophys.*, **332**, 280–287.

13.   THE CHEMISTRY AND APPLICATIONS OF ANTHOCYANINS AND FLAVONES FROM PERILLA LEAVES

KUMI YOSHIDA[1], KIYOSHI KAMEDA[1] and TADAO KONDO[2]

[1]*School of Life Studies, Sugiyama Jogakuen University, 17-3 Hoshigaokamotomachi, Chikusa, Nagoya 464, Japan,*
[2]*Chemical Instrument Center, Nagoya University, Furo-cho, Chikusa, Nagoya 464-01, Japan*

INTRODUCTION

Anthocyanins are contained in red, purple, and blue colored flowers, fruits, leaves, and roots of higher plants. They mainly exist as glycosides in plants and their aglycon, anthocyanidin, is a chromophore in pigments (Timberlake and Bridle, 1975; Timberlake and Bridle, 1982; Harborne and Grayer, 1988; Goto and Kondo, 1991; Mazza and Miniati, 1993; Strack and Wray, 1994). Anthocyanins change their color with pH, like litmus (Willstätter and Everest, 1913) and, as shown in Figure 1, form flavylium ions in strongly acidic solutions resulting in a very stable orange to red. In weakly acidic or neutral solutions they first begin to form anhydrobases, so that the color is reddish violet to violet. The blue they produce in alkaline media is the predominance of anhydrobase anions. However, anhydrobases and anhydrobase anions are unstable and are easily hydrated at the 2-position of the anthocyanidin nucleus, resulting in a rapid change to the colorless pseudobase (Brouillard and Delaporte, 1977; Brouillard, 1982).

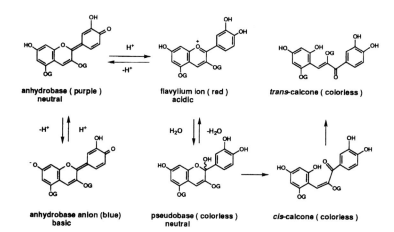

anhydrobase (purple)
neutral

flavylium ion (red)
acidic

trans-calcone (colorless)

anhydrobase anion (blue)
basic

pseudobase (colorless)
neutral

cis-calcone (colorless)

Figure 1   Structural changes of anthocyanins with pH in aqueous solutions

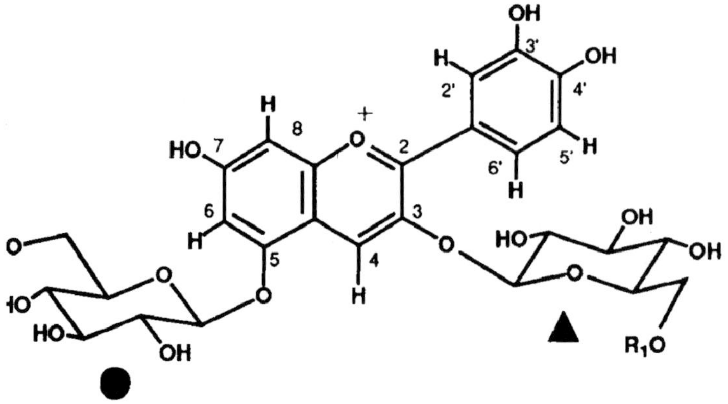

		R$_1$	R$_2$
malonylshisonin	(1)	E-p-coumaroyl	malonyl
shisonin	(2)	E-p-coumaroyl	H
caffeylmalonylcyanin	(3)	E-caffeoyl	malonyl
malonyl-*cis*-shisonin	(4)	Z-p-coumaroyl	malonyl
caffeylcanin	(5)	E-caffeoyl	H
cis-shisonin	(6)	Z-p-coumaroyl	H
cyanin	(7)	H	H

Figure 2 Structures of *Perilla* anthocyanins

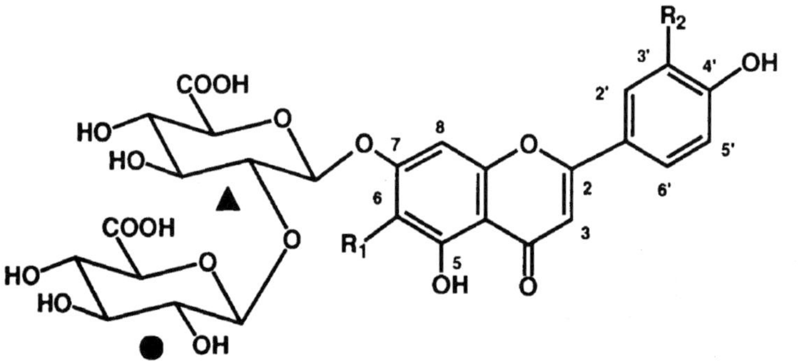

		R1	R2
7-O-diglucuronylapigenin	(8)	H	H
7-O-diglucuronylluteolin	(9)	H	OH
7-O-diglucuronylscutellarein	(10)	OH	H

Figure 3 Structures of *Perilla* flavones

Purple leaves of *Perilla ocimoides* recently renamed as *P. frutescens* (Japanese name, akashiso or akachirimenshiso) have been traditionally used for coloring Japanese pickles. The red color of pickled plum, Umeboshi, and pickled ginger, Benishoga, is achieved with the aid of the anthocyanins in Perilla leaves. Therefore, in Japan the characteristics of Perilla anthocyanins and flavones have attracted attention for a long time. Kondo (1931) and Kuroda and Wada (1935) isolated an anthocyanin pigment from purple Perilla leaves and gave it the name shisonin. Takeda and Hayashi (1964) and Watanabe *et al.* (1966) separately determined the structure of shisonin to be cyanidin 3-(6-*p*-coumaryl-D-glucoside)-5-glucoside, and this was confirmed and extended by ourselves (Goto *et al.*, 1978). Ishikura (1981) reported several kinds of anthocyanins and flavones in Perilla leaves and the complete structures of the major and minor anthocyanins from purple leaves were determined by Kondo *et al.* (1989) and Yoshida *et al.* (1990).

Recently interest in anthocyanins has increased, since it is a safe food colorant, in contrast to the synthetic azo dyes which often possess the potential for toxicity (Drake, 1980). Furthermore they can be also expected to have antioxidant activity (Tsuda *et al.*, 1994) and a cholesterol lowering action (Igarashi *et al.*, 1990; Igarashi and Inagaki, 1991). Perilla leaf extract is already commercially available as a red food colorant. However, several problems remain regarding usage i. e. color stability and color control. In this chapter we will cover the flavonoid components of purple Perilla leaves from the chemical view point (Kondo *et al.*, 1989; Yoshida *et al.*, 1990; Yoshida *et al.*, 1993a). We will describe the analytical methods used for the examination of anthocyanins and flavones, along with procedures for their isolation and elucidation of their chemical structures. We also describe chemical studies of their stability and color variation (Yoshida *et al.*, 1993b).

ISOLATION AND HPLC ANALYSIS OF ANTHOCYANINS AND FLAVONES

Anthocyanins and flavones have been analyzed using paper chromatography, several kinds of low pressure column chromatography and thin-layer chromatography (Markham, 1975; Francis, 1982). From the end of the 70's, the HPLC technique has developed remarkably and its application for the analysis of anthocyanins and flavones has markedly progressed (Adamovics and Sterimitz, 1976; Wilkinson *et al.*, 1977; Williams *et al.*, 1978; Wulf and Nagel, 1978; Asen, 1979; Akavia and Strack, 1980a; Strack *et al.*, 1980b). There is a long list of reviews of the methods now available (Hostettmann and Hostettmann, 1982; Francis, 1982; Harborne and Grayer, 1988; Daigle and Conkerton, 1988). From the middle of the 80's, the introduction of photo-diode array detectors facilitated identification of anthocyanins and flavones in plant extracts (Andersen, 1985; Idaka *et al.*, 1987; Andersen, 1987; Law and Das, 1987; Oleszek *et al.*, 1988; Daigle and Conkerton, 1988; Hebrero *et al.*, 1989; Hong and Wrolstad, 1990a; Yoshida *et al.*, 1990; Yoshida, 1992; Strack and Wray, 1994).

Using HPLC analysis it was revealed that the traditional isolation procedure with HCl-methanol solution was not particularly appropriate. When anthocyanins and flavones have malonyl resides or glucuronic acid resides in their molecules (Figure 2, 3), these are easily removed or modified (esterified at carboxyl groups or hydrolized) during extraction

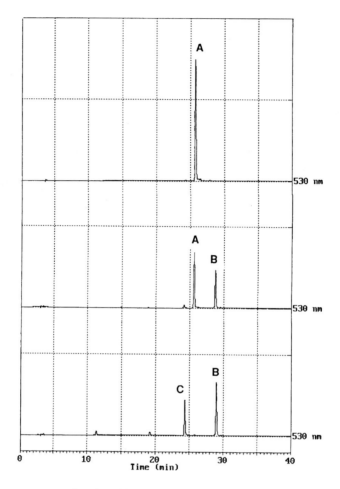

Figure 4 Chromatogram of malonylshisonin (**1**) using 1% HCl -methanol. The upper figure: analysis just after dissolving, A=(**1**); middle figure: analysis after 1 hour at room temperature, B=methyl ester of malonylshisonin; lower figure, analysis after 1 day at room temperature, C=(**2**).

and purification (Goto *et al.*, 1983; Tamura *et al.*, 1983; Kondo *et al.*, 1989). Figure 4 illustrates the degradation process of the major anthocyanin, malonylshisonin (**1**, peak A), isolated from the leaves. Under the given conditions a new peak (peak B) appeared after 1 hour at room temperature, and after 24 hours treatment peak A disappeared to become converted to a new peak (peak C) with a shorter retention time. In this case the structure of peak B pigment is methyl ester of malonylshisonin, that of peak C pigment is shisonin (**2**). Thus, one should not use alcoholic solvents for extraction of anthocyanins and flavones before elucidating their structures to confirm that they do not possess unstable residues. Aqueous trifluoroacetic acid (TFA) solution or the same solution with the addition of 50% of acetonitrile is generally more suitable.

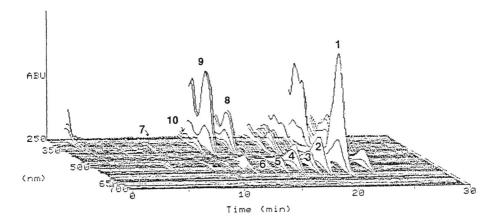

Figure 5   Chromatogram of an extract from *perilla ocimoide* using 3% TFA, as detected by phoro-diode array. (Partially revised from Yoshida *et al.* in 1990 with permission.)

A HPLC chromatogram of an extract of Perilla leaves, using a photodiode array detector is shown in Figure 5 (Yoshida *et al.*, 1990). It is clear that the purple colored leaves contain more than ten flavonoid compounds. From the UV/VIS spectra we can elucidate whether each anthocyanin component has an aromatic acid residue or not. The ratio of the maximum absorbances for the UV region (from 260 nm to 350 nm, E_{uvmax}) and the visible region (from 500 nm to 600 nm, E_{vismax}) indicates the number of the cinnamic acid derivative residues in the molecule (Harborne, 1958; Idaka *et al.*, 1987; Hong and Wrolstad, 1990b; Yoshida, 1992). According to this experimental rule the index for major anthocyanin from Perilla was calculated to be about 0.7, then it was supposed to possess one aromatic acyl residue.

EXTRACTION AND PURIFICATION OF ANTHOCYANINS AND FLAVONES

To avoid degradation, acidic water and/or acidic water-acetonitrile solution are used for the extraction of anthocyanins. However, these solvents are less able to permeate into plant cells than HCl-methanol solution (anthocyanins and flavones exist in vacuoles). In order to increase extraction yield fresh plant material is therefore generally first frozen in liquid nitrogen and pulverized in a blender or in a mortar, then extracted (Goto *et al.*, 1984). To extract Perilla flavonoids we adopted the same procedure (Kondo *et al.*, 1989; Yoshida *et al.*, 1990).

For purification, mass paper chromatography, cellulose powder column chromatography, polyvinylpyrolidone column chromatography and ion exchange column chromatography were earlier used (Markham, 1975; Timberlake and Bridle, 1975). The disadvantage is that these columns absorb fairly high amounts of anthocyanins and flavones irreversibly. After testing more than 30 kinds of chromatographic gels we found

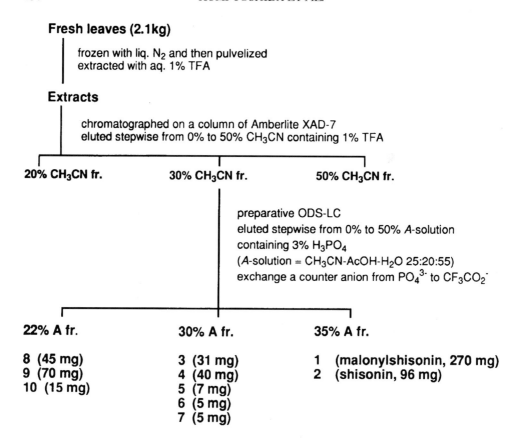

Figure 6   Isolation procedure of anthocyanins and flavones from purple leaves of *Perilla ocimoides*

the Amberlite XAD-7 gel, which has both hydrophobic and weak cation exchange properties, to be the most suitable for purification of anthocyanins (Goto *et al.*, 1982). The raw extract was evaporated under reduced pressure and the acetonitrile was removed. The resulting extract was poured into an Amberlite XAD-7 column, which was first washed with aqueous TFA solution to elute polysaccharides and other polar contaminants, then eluted stepwise by using aq. acetonitrile solution containing TFA. Anthocyanin containing fractions and flavone containing fractions were collected and evaporated under reduced pressure below 40°C (Figure 6).

Crude pigment was then purified using preparative reversed phase column chromatography as shown in Figure 6. For this ODS-LC the column was eluted stepwise with aqueous *A*-solution containing 3% phosphate; *A*-solution:acetonitrile:acetic acid: water = 25:20:55 (Akavia and Strack, 1980a). Highly endcapped ODS columns allow the used acetic acid to be removed from the elution solution and therefore aqueous TFA-acetonitrile solution can be used. Applying this method we were able to isolate 270 mg of the major anthocyanin (malonylshisonin) and about 184 mg of minor anthocyanins, as well as 130 mg of flavones from 2.1 kg of fresh purple Perilla leaves.

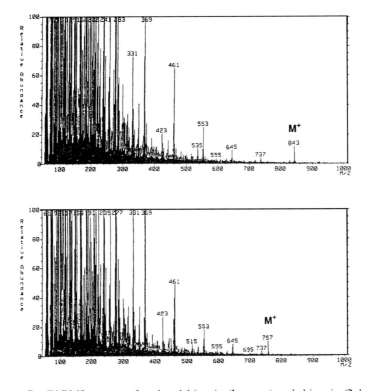

Figure 7 FABMS spectra of malonylshisonin (**1**, upper) and shisonin (**2**, lower)

STRUCTURAL DETERMINATION OF ANTHOCYANINS AND FLAVONES

The structure of shisonin isolated from fresh purple leaves, *Perilla ocimoides,* was first elucidated to be *p*-coumaroylcyanin by ordinary acid hydrolysis (Kuroda and Wada, 1935a, b), and then determined by Karrer's hydrogen peroxide method (Karrer and Widmer, 1927) to be cyanidin 3-(6-*p*-coumaryl-D-glucosido)-5-glucoside (Takeda and Hayashi, 1964; Watanabe, 1966). For the complete structure with the stereo configuration to be ascertained a non-degrading method had to be developed. Now the structures of flavonoids can be determined completely using instrumental analysis (MS, NMR). The structures of Perilla anthocyanins and flavones elucidated in this way are shown in Figure 2 and 3.

The molecular weights of anthocyanins are determined by positive FAB-MS (fast atom bombardment mass spectroscopy) using a strongly acidic matrix such as HCl-glycerol and *A*-solution-*m*-nitrobenzylalcohol. The fragment ions also provide structural information. By using FABMS (Figure 7) the molecular ion peaks of malonlylshisonin (**1**) and shisonin (**2**) were found to be *m/z* 843 and 757, respectively, with the typical differentiation showing the existence of a malonyl group in (**1**). In fact acid methanolysis of **1** gives **2** and dimethylmalonate. Analysis of other minor pigments and flavones by FABMS gives molecular ion peaks.

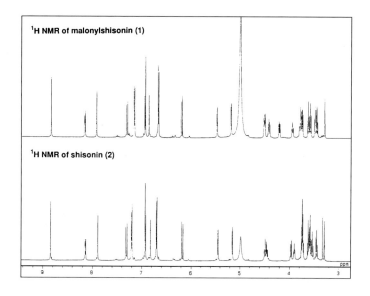

Figure 8   ¹H NMR spectra of malonylshisonin (**1**, upper) and shisonin (**2**, lower) in 10% TFA*d*-CD₃OD at 25°C (600 MHz)

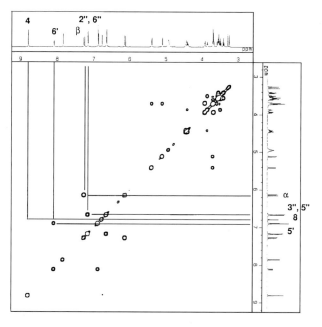

Figure 9   ¹H-¹H COSY spectrum of shisonin (**2**) in 10% TFA*d*-CD₃OD at 25°C (600 MHz)

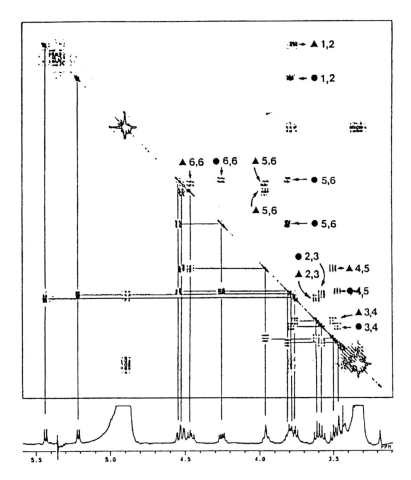

Figure 10   ¹H-¹H COSY spectrum of the sugar moieties of shisonin (**2**) in 3% TFA*d*-CD₃OD at 25°C (500 MHz, From Kondo *et al.* in 1989 with permission.)

The first NMR measurement of anthocyanins including shisonin was achieved (Goto *et al.*, 1978) in strongly acidic media. Under these conditions, anthocyanin exists as a single species to give a highly qualified spectrum, whereas in plants under physiological conditions, which are weakly acidic and neutral, equilibrium mixtures are found. For NMR measurement CD₃OD-DCl, CD₃OD-TFA*d*, and DMSO*d*6-TFA*d* have been used. But in CD₃OD -DCl demalonylation often occurred in our experience. In Figure 8 the ¹H NMR of **2** in CD₃OD -TFA*d* showed signals attributable to one cyanidin, two hexoses and one (*E*)-*p*-coumaric acid ($J_{\alpha,\beta}$=16Hz). With ¹H-¹H COSY (correlated spectroscopy, Figure 9) of **2**, low field signals could easily be assigned on the basis of a characteristic H-4 (Nilsson, 1973), and a correlated signal was attributable to H-8 (W-type long distance coupling). The COSY of sugar region revealed the correlations of the signals of the two hexoses (Figure 10). The vicinal coupling constants (*J*) of both sugars were found

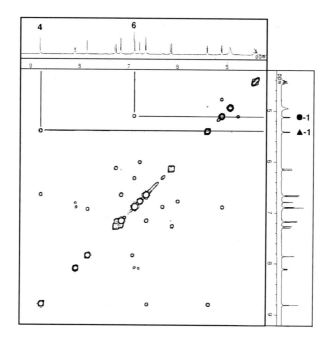

Figure 11   NOESY of shisonin (**2**) in 10% TFA*d*-CD$_3$OD at 25°C (600 MHz)

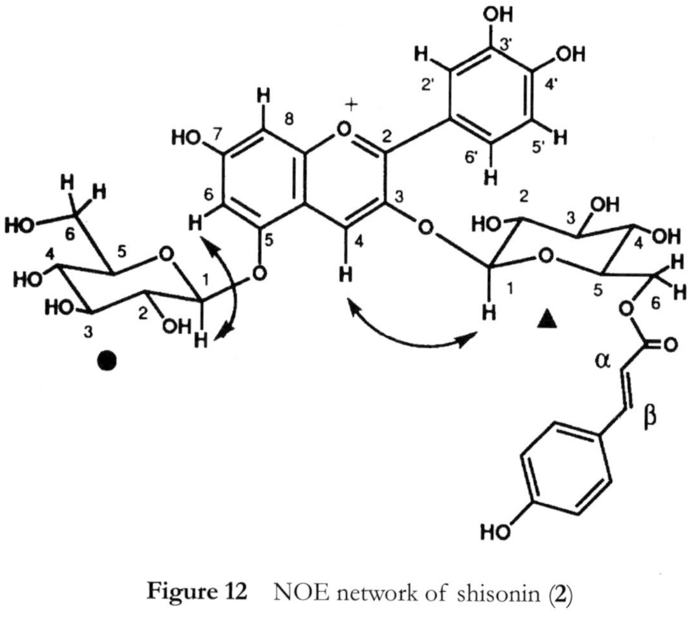

Figure 12   NOE network of shisonin (**2**)

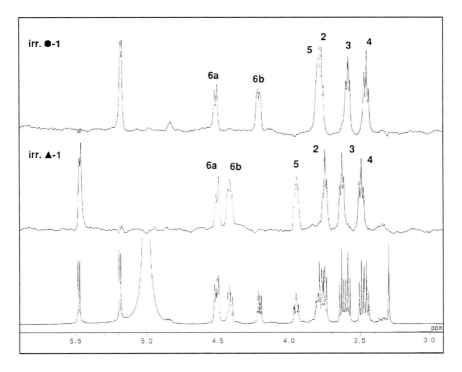

Figure 13   1D HOHAHA of the sugar region of malonylshisonin (**1**) irradiated at the anomeric protons (10% TFA*d*-CD$_3$OD at 25°C, 600 MHz)

to be in the range of 7.5–9.5 Hz, indicating that the hexosides were β-D-glucopyranosides. To clarify the linkage position among the residues NOE (nuclear Overhauser effect) difference spectra and/or NOESY (nuclear Overhauser effect spectroscopy) were recorded (Figure 11). The negative NOE observation between the anomeric proton and the ring proton of the cyanidin nucleus indicated the linkage position (Figure 12). The NOEs between ▲-1 and H-4, and ●-1 and H-6, established that the ▲ and ●-glucoses are attached at the 3 and 5 positions of the cyanidin nucleus, respectively. Finally, the *p*-coumaryl group was attached at 6-OH of the ▲-glucose of whose 6-Hs appeared on a 0.5 ppm lower field shift. Therefore, the structure of **2** was determined to be 3-*O*-(6-*O*-(*E*)-*p*-coumaryl-β-D-glucopyranosyl)-5-*O*-(β-D-glucopyranosyl)cyanidin. In the case of **1**, most of the signals were very similar to those for **2** (Figure 8) and could be assigned using the same procedure. 1D HOHAHA (homonuclear Hartman Hahn experiment, Davis and Bax, 1985) irradiatiting at the anomeric proton revealed the signals for the individual sugars (Figure 13). The difference between the spectra of **1** and **2** was only the signals of sugars (methylene signals of malonyl group were absent arising from replacement of deuterium). H-6 of the ● -glucose of **1** was shifted toward the lower field, revealing a structure for **1** as 3-*O*-(6-*O*-(*E*)-*p*-coumaryl-β-D-glucopyranosyl)-5-*O*-(6-*O*-malonyl-β-D-glucopyranosyl) cyanidin.

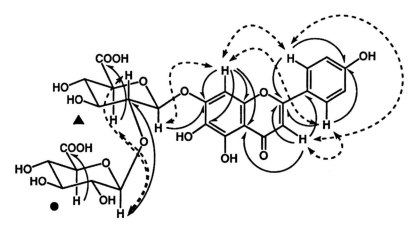

Figure 14   NOE (dotted lined double headed arrow) and HMBC (solid lined double headed arrow) correlations of apigenin diglucuronide (**8**)

Structure of the other minor anthocyanins has similarly been determined as shown in Figure 2. In Perilla leaves, new type anthocyanins containing *Z-p*-coumaric acid (**4** and **6**) were first found in this work (Yoshida *et al.*, 1990). ^{1}H NMR data for **4** and **6** were very similar to those for **1** and **2** except for the $J_{\alpha\beta}$ (12 Hz) indicating a *cis* (Z) configuration of the double bond in the cinnamic acid residue. Thus, **4** was defined as 3-O-(6-O-(Z)-*p*-coumaryl-β-D-glucopyranosyl)-5-O-(6-O-malonyl-β-D-gluco-pyranosyl)cyanidin and **6** as 3-O-(6-O-(Z)-*p*-coumaryl-β-D-glucopyranosyl)-5-O-(β-D-glucopyranosyl)cyanidin.

In addition to the anthocyanins, the existence of three flavones in Perilla (Figure 5) were demonstrated, two (**8**, **9**) being already known and the other (**10**) a novel form (Figure 3, Yoshida *et al.*, 1993a). The UV data of **8**, **9**, and **10** indicated the presence of luteolin, apigenin and scutellarein chromophores, respectively, while their FABMS suggested the existence of two additional hexouronic acid residues. By ^{1}H-^{1}H COSY, HOHAHA, NOE difference, all the proton signals could be assigned. ^{1}H NMR of **8**, **9**, and **10** confirmed the presence of these components and furthermore the two hexouronic acids residues were determined to both be β-glucopyranosiduronic acid ($J_{1,2}$=7.5, $J_{2,3}$=$J_{3,4}$=$J_{4,5}$=9.0 Hz and H-5: doublet). The remaining problem was the attachment positions of the glucuronic acids. When H-1 of the ▲ and ●-glucuronic acids in **8** were irradiated, strong NOEs were observed to H-8, and the H-2 and H-3 of ▲, respectively (Figure 14). Thus the position of ▲ was determined to be the 7-OH of the flavone chromophore. However, whether the position of ● was hydroxy group at C-2 or C-3 of ▲ could not be determined. To elucidate the position of the ●-glucuronic acid, HMBC (heteronuclear multiple-bond connectivity, Bax and Summers, 1986) were recorded and all the ^{1}H and ^{13}C signals were assigned. In the HMBC, H-1s of ▲ and ● correlated with the C-7, and C-2 of ▲, respectively (Figure 15). Therefore, the linkage positions of all sugars could be determined and the structure of **8** was established to be 7-O-(2-O-(β-D-glucopyranosyluronic acid)-β-D-glucopyranosyluronic acid) apigenin. Compound **9** and **10** are 7-diglucuronyl luteolin and scutellarein, respectively.

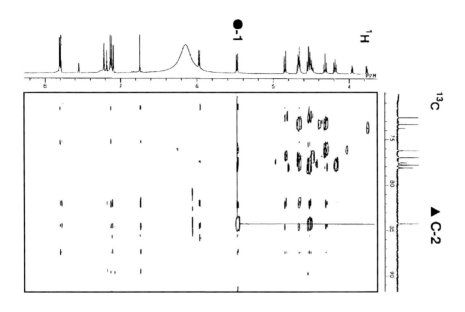

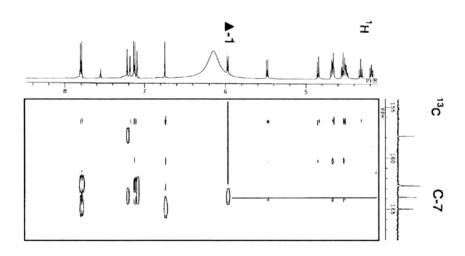

Figure 15   HMBC of apigenin diglucuronide (**8**) in pyridine-d_5 at 25°C (500 MHz)

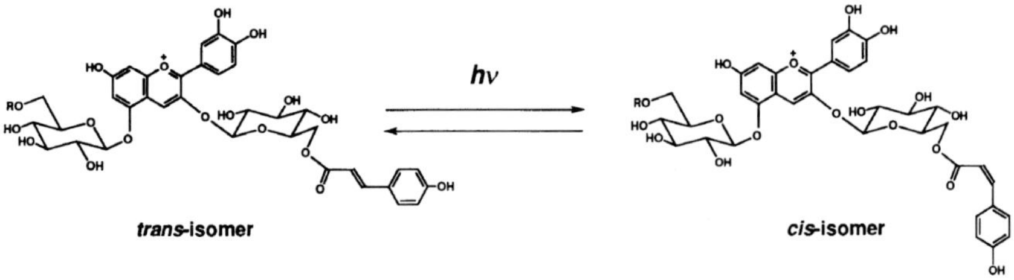

Figure 16 *Cis-trans* isomerization of malonylshisonin (**1**) by irradiation with light

CIS-TRANS ISOMERIZATION OF CINNAMIC ACID RESIDUES OF *PERILLA* ANTHOCYANINS BY LIGHT

Many acylated anthocyanins which contain cinnamic acid derivatives, i. e. *p*-coumaric acid, caffeic acid, and feruric acid etc., have been isolated. However, no derivatives with a *cis*-configuration of the α–β double bond position were found until Yoshida *et al.* (1990) reported the structure of *cis*-malonylshisonin (**4**) and other anthocyanins with *cis*-cinnamic acid residues in purple Perilla leaves. Subsequently, *cis*-malonylawobanin was also isolated from petals of the blue day flower, *Commelina communis* (Kondo *et al.*, 1991). The presence of *cis*-isomers of cinnamic acid residues of anthocyanins gave rise to the question of whether *cis-trans* isomerization (Figure 16) might occur with exposure to daylight. At the same time it was questioned why only monoacylated anthocyanins have *cis*-isomers and no polyacylated anthocyanins share this feature. These problems have attracted a great deal of interest from the view point of the light-resistance of anthocyanin pigments not only in the plant body, but also when used for coloring purposes.

The *cis-trans* isomerization of malonylshisonin and *cis*-malonylshisonin was examined by Yoshida *et al.* (1990). Acidic methanol solutions of malonylshisonin and *cis*-malonylshisonin were irradiated by high-pressure Hg lamp (366 nm) at room temperature. With both compounds the isomerization reaction reached equilibrium within 30 min and the ratio of *cis*- and *trains*-isomers was 1:1 (Figure 17). In acidic aqueous solutions the same isomerization reaction occurred, although the equilibrium ratio of the *cis*-isomer to the *trans*-isomer was then 1:4. In neutral aqueous solutions the equilibrium ratio again differed, being 3:7 for *cis*- to *trans*-. Not only malonylshisonin but also other anthocyanins which contain cinnamic acid residues isomerized when irradiated with UV light (Yoshida *et al.*, 1990). Since the same reaction occurred with exposure to sun light, *cis*-isomers may also be produced in plant leaves under natural conditions.

The energy of strong UV light is very destructive for some components of plant leaves and petals and much damage occurs due to sun light. For example, the chlorophyll in chloroplasts is continuously broken down and replaced. The coexistence of carotenoids is very important to reduce the amounts of free radicals and active oxygen species. The isomerization reaction in response to light may also have the effect of quenching the high energy of short wavelength light. In future, the function of anthocyanins in plant cells should therefore be investigated from the view point of the light-resistance.

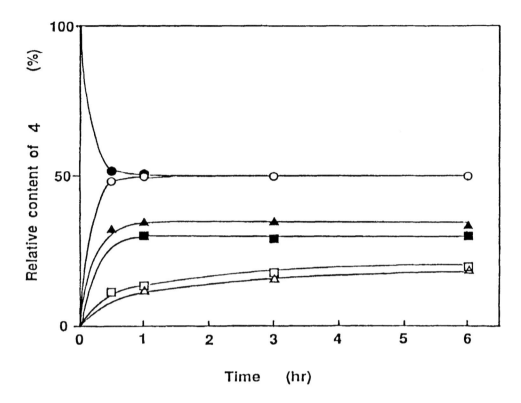

Figure 17 Relative content of the isomerization product, *cis*-malonylshisonin (**4**), caused by light. ○ from **1** in methanol (10^{-2} M) irradiated by UV;
□ from **1** in the buffer (10^{-2} M, pH 2.0); irradiated by UV;
■ from **1** in the buffer (10^{-3} M, pH 6.0); irradiated by UV;
Δ from **1** in the buffer (10^{-2} M, pH 2.0); irradiated by sunlight;
▲ from **1** in the buffer (10^{-3} M, pH 6.0) irradiated by sunlight;
● from **4** in methanol (10^{-2} M); irradiated by UV. (from Yoshida *et al.* in 1990 with permission.)

STABILITY OF *PERILLA* ANTHOCYANINS UNDER HIGHLY ACIDIC AND SALT CONDITIONS

The red color of Umeboshi, a Japanese traditional pickle, is stable with storage for several years and this stability might be connected with the low pH and high salt concentration. Goto *et al.* (1976) reported that anthocyanins can be stabilized with 4 M NaCl and 4 M $MgCl_2$ after extracting the genuine anthocyanins from pansy petals with 4 M $MgCl_2$. Takeda *et al.* (1985) also succeeded in extracting of blue colored pigments with 4 M NaCl from blue sepals of *Hydrangea macrophylla*. The effects of metal ions in food were reviewed by Markakis (1982).

Table 1 λ_{vismax} (nm) of malonylshisonin (**1**) in various salt solutions. $(5 \times 10^{-5}M, pH\ 6)$

		Conc of salt	
		$5 \times 10^{-3}M$	$1M$
buffer	542	-	-
NaCl	-	542	548
LiCl	-	542	543
KCl	-	542	546
MgCl$_2$	-	541	581
CaCl$_2$	-	542	583
FeCl$_3$		549	-
AlCl$_3$	-	543	-

Yoshida *et al.* (1993b) investigated the color variation and stability of Perilla anthocyanins in various inorganic salt solutions. In buffer solution without salt, malonylshisonin (**1**) showed a purple color and the absorption maximum (λ_{max}) at the visible region was 542 nm (Table 1). In 1 M aqueous salt solutions, monovalent inorganic salts did not have any obvious bathochromic effect but divalent salts, MgCl$_2$ and CaCl$_2$, shifted the λ_{vismax} to about a 40 nm longer wavelength. In neutral solutions it appears that a concentrated divalent metal ion could be co-ordinated with the *ortho*-dihydroxyl group of the B-ring of the anthocyanidin nucleus (Bayer, 1966; Hayashi and Takeda, 1970; Kondo *et al.*, 1992; Kondo *et al.*, 1994).

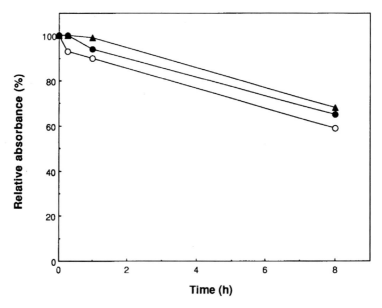

Figure 18   Color stability of shisonin (**2**) in concentrated NaCl solutions. (shisonin; 5×10^{-5} M, pH 2.0, 20°C). Stability was indicated by the relative absorbance at the λ_{vismax} of each solution. In buffer (○), In 1 M NaCl solution (●), in 4 M NaCl solution (▲)

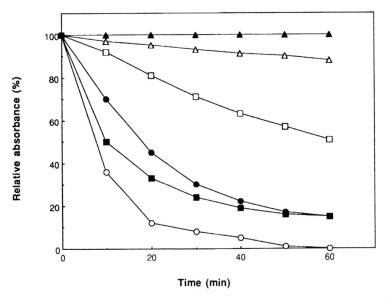

Figure 19   Color stability of shisonin (**2**) in concentrated salt solutions. (shisonin; 5×10^{-5} M, pH 6.0, 20°C). Stability was indicated by the relative absorbance at the λ_{vismax} of the each solution. In buffer (O), In 4 M NaCl solution (●), in 4 M LiCl solution (□), in 4 M KCl solution (■), in 1 M MgCl$_2$ solution (△), in 4 M MgCl$_2$ solution (▲)

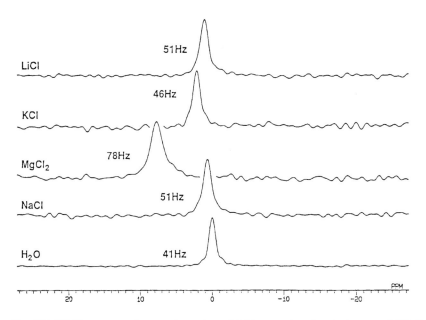

Figure 20 ^{17}O NMR spectra of water in various 4 M inorganic salt solutions at 40°C (36.58 MHz, ^{1}H decoupled)

In strongly acidic solutions the color of shisonin (**2**) is very stable and the concentration of NaCl solution does not show remarkable effects (Figure 18). In neutral solutions, however, the color is very unstable and is quickly lost. This can be blocked by addition of concentrated divalent salt, $MgCl_2$ showing the strongest stabilizing effect (Figure 19). The decolorization of anthocyanin is caused by water attack at the 2-position of anthocyanidin nucleus (Brouillard and Delaporte, 1977; Brouillard, 1982) and in concentrated salt solutions water mobilizing activity would be expected to be reduced, so that the hydration reaction might be prevented (Goto *et al.*, 1976). To evaluate water mobility, the ^{17}O NMR (Richardson and Steinberg, 1987) of high concentrated salt solutions was measured (Yoshida *et al.*, 1993b) and as shown in Figure 20, the line width of 4 M $MgCl_2$ was found to be the widest followed by that of 4 M LiCl. The line width of ^{17}O signal of salt solutions in fact correlates with the stabilizing effect, suggesting that suppression of water mobility does indeed lead to stabilization of pigments.

SUMMARY

In purple leaves of *Perilla ocimoides* several anthocyanins and flavones exist. Using TFA-acetonitrile solution they could be extracted without any degradation. Each pigment and flavone was isolated by using Amberlite XAD-7 column chromatography followed by ODS-HPLC. The structure of them was elucidated by chemical instrumental analysis i. e. FABMS and NMR. With FABMS using a strongly acidic matrix the exact molecular weight of anthocyanins could be directly determined. Various 1D and 2D measurement of NMR, COSY, HOHAHA, NOESY, HMBC etc. clarified the structure of anthocyanidin nucleus, the position of the sugars and acids, and the geometry of double bond of cinnamic acid derivatives. The true structure of the major anthocyanin in purple Perilla leaves was malonylshisonin. In the minor anthocyanins *cis*-cinnamic acid residues were first observed and *cis-trans* isomerization by irradiation of light was also studied. Perilla anthocyanins were stabilized in strong acidic solutions or in highly concentrated salt solutions. The suppression of water mobility may prevent the hydration reaction of anthocyanidin nucleus. Perilla anthocyanins are expected as a source of multi-functional food colorant for light preventing effect, antioxidant activity and a cholesterol lowering action.

REFERENCES

Adamovics, J. and Stermitz, F.R. (1976) High-performance Liquid Chromatography of Some Anthocyanidins and Flavonoids. *J. Chromatography*, **129**, 464–465.

Akavia, N. and Strack, D. (1980a) High Performance Liquid Chromatography of Anthocyanidins as a New Approach to Study Flower Pigment Genetics. *Z. Naturforsch.*, **35C**, 16–19.

Andersen, O.M. (1985) Chromatographic Separation of Anthcyanins in Cowberry (Lingonberry) *Vaccinium vites-ideaea* L. *J. Food. Sci.* **50**, 1230–1232.

Andersen, O.M. (1987) Anthocyanins in Fruits of *Vaccinium uloginosum* L. (Bog Whortleberry). *J. Food. Sci.* **52**, 665–666.

Asen, S. (1979) Flavonoid Chemical Markers in Poinsettia Bracts. *J. Amer. Soc. Hort. Sci.*, **104**, 223–226.

Bayer, E. (1966) Complex Formation and Flower Colors. *Angew. Chem. Int. Ed. Engl.*, **5**, 791-798.

Bax, A. and Summers, M. F. (1986) ^{1}H and ^{13}C Assignments from Sensitivity-Enhanced Detection of Heteronuclear Multiple-Bond Connectivity by 2D Multiple Quantum NMR. *J. Amer. Chem. Soc.*, **108**, 2093–2094.

Brouillard, R and Delaporte, B. (1977) Chemistry of Anthocyanin Pigment. 2. Kinetic and Thermodynamic Study of Proton Transfer, Hydration, and Tautomeric Reactions of Malvin 3-Glucoside. *J. Amer. Chem. Soc.*, **99**, 8461–8468.

Brouillard, R. (1982) Chemical Structure of Anthocyanins. In Markakis, P. (eds.), *Anthocyanins as Food Colors*, Academic Press, New York, pp. 1–40.

Daigle, D.J. and Conkerton, E.J. (1988) Analysis of Flavonoids by HPLC: An Update. *J. Liquid Chromatography*, **11**, 309–325.

Davis, D. G. and Bax, A. (1985) Assignment of Complex ^{1}H NMR Spectra via Two-Dimensional Homonuclear Hartmann-Hahn Spectroscopy. *J. Amer. Chem. Soc.* **107**, 2820–2821.

Drake, J.J-P. (1980) In "Developments in Food Colours-1" ed. by J. Walford, Applied Science. Publishers, London, pp. 219–253.

Francis, F.J. (1982) Analysis of Anthocyanins. In Markakis, P. (eds.), *Anthocyanins as Food Colors*, Academic Press, New York, pp. 182–207.

Goto, T., Hoshino, T. and Ohba, M. (1976) Stabilization effect of neutral salts on anthocyanins, flavylium salts, anhydrobases and genuine anthocyanins. *Agric. Biol. Chem.*, **40**, 1593–1596.

Goto, T., Takase, S. and Kondo, T. (1978) PMR Spectra of Natural Acylated Anthocyanins: Determination of Stereostructure of Awobanin, Shisnonin and Violanin. *Tetrahedron Lett.*, **30**, 420–423.

Goto, T., Kondo, T., Tamura, H. and Imagawa, H. (1982) Structure of Gentiodelphin, an Acylated Anthocyanin Isolated from *Gentiana makinoi*, That is Stable in Diluted Aqueous Solution. *Tetrahedron Lett.*, **23**, 3695–3698.

Goto, T., Kondo, T., Tamura, H. and Takase, S. (1983) Structure of Malonylawobanin, the Real Anthocyanin Present in Blue-colored Flower Petals of *Commelina communis. Tetrahedron Lett.*, **24**, 4863–4866.

Goto, T., Kondo, T., Kawai, T. and Tamura, H. (1984) Structure of Cinerarin, a Tetra-acylated Anthocyanin Isolated form the Blue Garden Cineraria, *Senecio cruentus.Tetrahedron Lett.*, **25**, 6021–6024.

Goto, T. and Kondo, T. (1991) Structure and Molecular Stacking of Anthocyanins—Flower Color Variation. *Angew. Chem. Int. Engl.* **30** 17–33.

Harborne, J.B. (1958) Spectral Methods of Characterizing Anthocyanins. *Biochem. J.*, **70**, 22–28.

Harborne, J.B. and Grayer, R.J. (1988) The Anthocyanins. in J.B. Harborne, (eds.), *The Flavonoids: Advances in Research since 1980*, Chapman and Hall, London, pp. 1–20.

Hayashi, K. and Takeda, K. (1970) Further Purification and Component Analysis of Commelinin Showing the Presence of Magnesium in This Blue Complex Molecule. *Proc. Jpn. Acad.*, **46**, 535–540.

Hebrero, E., Gracia-Rodriguez, C., Santos-Buelga, C. and Rivas-Gonzalo, J.C. (1989) Analysis of Anthocyanins by High Performance Liquid Chromatography-Diode Array Spectroscopy in a Hybrid Grape Variety (*Vitis vinifera* x *Vitis berlandieri* 41B). *Am. J. Enol. Vitic.*, **40**, 283–291.

Hong, V. and Wrolstad, R.E. (1990a) Characterization of Anthocyanin-Containing Colorants and Fruit Juices by HPLC/Photodiode Array Detection. *J. Agric. Food Chem.*, **38**, 698–708.

Hong, V. and Wrolstad, R.E. (1990b) Use of HPLC separation/Photodiode Array Detection for Characterization of Anthocyanins. *J. Agric. Food Chem.*, **38**, 708–715.

Hostettmann, K. and Hostettmann, M. (1982) Isolation Techniques for Flavonoids. In Harborne, J.B., Mabry, T.J. (eds.), *The Flavonoids: Advances in Research*, Chapman and Hall, London, pp. 1–18.

Idaka, E., Ogawa, T., Kondo, T. and Goto, T. (1987) Isolation of Highly Acylated Anthocyanins form *Commelinacea* Plants, *Zebrina pendula, Rhoeo spathacea* and *Sercreasea purpurea. Agric. Biol. Chem.*, **51**, 2215–2220.

Igarashi, K., Abe, S. and Satoh. J. (1990) *Agric, Biol. Chem.*, **54**, 171–175.

Igarashi, K, and Inagaki, K. (1991) Effects of the Major Anthocyanin of Wild Grape (Vitis coignetiae) on Serum Lipid Levels in Rats. *Agric, Biol. Chem.*, **55**, 285–287.

Ishikura, N. (1981) Anthocyanins and Flavones in Leaves and Seeds of *Perilla* plant. *Agric. Biol. Chem.*, **45**, 1855–1860.

Karrer, P. and Widmer, R. (1927) Über Pflanzenfarbstoffe II. *Helv. Chim. Acta*, **10**, 67–86.

Kondo, K. (1931) Untersuchung über Anthocyanin und Anthocyanidin (V.) über den Farbstoff von *Perilla ocimoides* L. var. *crispa* Benth. *Yakugaku Zasshi*, **51**, 254–260.

Kondo, T., Tamura, H., Yoshida, K. and Goto, T. (1989) Structure of Malonylshisonin, a Genuine Pigment in Purple Leaves of *Perilla ocimoides* L. var. *crispa* Benth. *Agric. Biol. Chem.*, **53**, 797–800.

Kondo, T., Yoshida, K. Yoshikane, M. and Goto, T. (1991) Mechanism for Color Development in Purple Flower of *Commelina communis. Agric. Biol. Chem.*, **55**, 2919–2921.

Kondo, T., Yoshida, K., Nakagawa, A., Kawai, T., Tamura, H. and Goto, T. (1992) Structural Basis of Blue-Colour Development in Flower Petals from *Commelina communis. Nature*, **358**, 515–518.

Kondo, T., Ueda, M., Tamura, H., Yoshida, K., Isobe, M. and Goto, T. (1994) Composition of Protocyanin, a Self-Assembled Supramolecular Pigment from the Blue Cornflower, *Centaurea cyanus. Angew. Chem. Int. Ed. Engl.*, **33**, 978–979.

Kuroda, C. and Wada, M. (1935a) The coloring Matter of Shiso. *Proc. Imp. Acad. Japan*, **11**, 28–31.

Kuroda, C. and Wada, M. (1935b) The Constitution of Natural Coloring Matters, Kuromamin, Shisonin, and Nasunin. *Proc. Imp. Acad. Japan*, **11**, 272–287.

Law, K.H. and Das, N.P. (1987) Dual-wavelength Absorbance Ratio and Spectrum Scaning Tecniques for Identification of Flavonoids by High-performance Liquid Chromatography. *J. Chromatography*, **388**, 225–233.

Markakis, P. (1982) Stabillity of Anthocyanins in Food. In Markakis, P. (eds.), *Anthocyanins as Food Colors*, Academic Press, New York, pp. 163–180.

Markham, K.R., (1975) Isolation Techniques for Flavonoids in J.B. Harborne, T.J. Mabry and H. Mabry, (eds), *The Flavonoids*, Chapman and Hall, London, pp. 1–44.

Mazza, G. and Miniati, E. (1993) *Anthocyanins in Fruits, Vegetables, and Grains*, CRC Press, Boca Raton.

Nilsson, E. (1973) Studies of Flavylium compounds VIII. Application of Proton NMR Spectroscopy to the Analysis of Anthocyanidin Pigments. *Chemica Scripta*, **4**, 49–55.

Oleszek, W., Lee, C.Y., Jaworski, A.W. and Price, K.R. (1988) Identification of Some Phenolic Compounds in Apples. *J. Agric. Food Chem.*, **36**, 430–432.

Richardson, S. and Steinberg, M.P. (1987) Application of Nuclear Magnetic Resonance. In Rockland, L.B. and Beuchat, L.R. (eds.), *Water Activity: Theory and Application to Food*, Maecel Dekker, New York, pp. 235–285.

Strack. D., Akavia, N. and Reznik, H. (1980b) High Performance Liquid Chromatographic Identification of Anthocyanins. *Z. Naturforsch.*, **35C**, 533–538.

Strack, D. and Wray, V. (1994) The Anthocyanins. In Harborne, J.B. (eds.), *The Flavonoids: Advances in Research Since 1986*, Chapman and Hall, London, pp. 1–22.

Takeda, K. and Hayashi, K. (1964) Oxidative Degradation of Acylated Anthocyanins Showing the Presence of Organic Acid-Sugar Linkage in the 3-Position of Anthocyanins; Experiments on Ensatin, Awobanin, and Shisonin. *Proc. Japan. Acad.*, **40**, 510–515.

Takeda, K., Kubota, R. and Yagioka, C. (1985) Copigment in the Blueing of Sepal Colour *Hydrangea macrophylla. Phytochemistry*, **24**, 1207–1209.

Tamura, H., Kondo, T., Kato, Y. and Goto, T. (1983) Structure of a Succinyl Anthocyanin and a Malonyl Flavone, Two Constituents of the Complex Blue Pigment of Cornflower *Centaurea cyanus*. *Tetrahedron Lett.*, **24**, 5749–5752.

Timberlake, C.F. and Bridle, P. (1975) Anthocyanins in J.B. Harborne, T.J. Mabry and H. Mabry, (eds), *The Flavonoids*, Chapman and Hall, London, pp. 214–266.

Timberlake, C.F. and Bridle, P., (1982) Distibution of Anthocyanins in Food Plants in P. Markakis, (eds.), *Anthocyanins as Food Colors*, Academic Press, New York, pp. 125–162.

Tsuda, T., Ohshima, K., Kawakishi, S. and Osawa, T. (1994) Antioxidative Pigments Isolated from the Seeds of *Phaseolus vulgaris* L. *J. Agric. Food Chem.*, **42**, 248–251.

Watanabe, S., Sakamura, S. and Obata, Y. (1966) The Structure of Acylated Anthocyanins in Eggplant and Perilla and the Position of Acylation. *Agric. Biol. Chem.*, **30**, 420–422.

Williams, M., Hrazdina, G., Wilkinson, M.M., Sweeny, J.G. and Iacobucci, G.A. (1978) High-Pressure Liquid Chromatographic Separation of 3-Glucosedes, 3, 5-Diglucosides, 3-(6-O-*p*-Coumaryl)glucosides and 3-(6-O-*p*-Coumarylglucoside)-5-glucosides of Anthocyanidins. *J. Chromatography*, **155**, 389-398.

Willstätter, R. and Everest, A.E. (1913) Untersuchungen über die Anthocyane; I. Über den Farbstoff der Kornblume. *Ann. Chemie*, **401**, 189–232.

Wilkinson, M., Sweeny, J.G. and Iacobucci, G.A. (1977) High-pressure Liquid Chromatography of Anthocyanidins. *J. Chromatography*, **132**, 349-351.

Wulf, L.W. and Nagel, C.W. (1978) High-pressure Liquid Chromatographic Separation of Anthocyanins of *Vitis vinifera*, *Am. J. Enol. Vitic.*, **29**, 42–49.

Yoshida, K., Kondo, T., Kameda, K. and Goto, T. (1990) Structures of Anthocyanins Isolated form Purple Leaves of *Perilla ocimoides* L. var. *crispa* Benth and Their Isomerization by Irradiation of Light. *Agric. Biol. Chem.*, **54**, 1745–1751.

Yoshida, K. (1992) Thesis (Nagoya University). pp. 11–24.

Yoshida, K., Kameda, K. and Kondo, T. (1993a) Diglucuronoflavones from Purple Leaves of *Perilla ocimoides*. *Phytochemistry*, **33**, 917–919.

Yoshida, K., Kameda, K., Kondo, T. and Goto, T. (1993b) Stabilization and Color Variation of anthocyanins with Inorganic Salts. T. Kakihana, T. Hoshi, (eds.), in *Seventh Symposium on Salt vol II*, Elsevier Science Publishers B. V., Amsterdam, pp. 623–630.

14. ANTHOCYANINS FROM PERILLA

LUCY SUN HWANG

Graduate Institute of Food Science and Technology
National Taiwan University, Taipei, Taiwan

CULTIVATION AND ANTHOCYANIN COMPOSITION OF PERILLA GROWN IN TAIWAN

Cultivation

Perilla (*Perilla frutescens* Brit.), which originated in China and India, is a popular aromatic condimental vegetable in Japan. Perilla is seldom consumed as a vegetable by the Chinese people; it is regarded as a medicinal plant. In Taiwan, large plantations of Perilla started in 1979. Japanese businessmen found that farmers in Tao Yuan Hsien, Kung-Kang Hsiang (in the north-western part of Taiwan) could produce Perilla which had better aroma and could also be harvested earlier than in Japan. Perilla has, thus, become an export vegetable for the Japanese market. Approximately 90% of the Perilla grown in Taiwan is exported to Japan.

Perilla is generally planted in February to March in Taiwan. The seeds which have been stored under refrigeration temperature can sprout in four to five days after sowing. It takes ten to fifteen days to sprout, otherwise. The most suitable sprouting temperature is 22–23°C. The Perilla plant grows slowly initially, and the best growing temperature is 20°C. In northern Taiwan, Perilla leaves can be harvested three months after the seedling stage. They are usually harvested from May to early August before flowering. In the past, Perilla leaves were picked by hand since only three centre leaves were harvested. Now, tea leaf harvesting machines are used in Taiwan to reduce the labour cost and speed up the harvesting process. In Kung-Kang Hsiang of Taiwan, one hectare of land can produce, on average, 20–30 tons of fresh Perilla leaves. In 1979, only 5 hectares of land were cultivated with Perilla in Kung-Kang. As the export of Perilla to Japan increased, the plantation area grew to nearly 30 hectares in 1990. This number declined to less than 25 hectares in 1995 due to the high cost of labour and land in Taiwan.

Anthocyanin Composition

Many varieties of Perilla are grown in Taiwan. Some are local wild types, and some were introduced from Japan. Their colours differ widely from green to purple. Our laboratory chose three purple varieties of Perilla to study the changes of anthocyanin content with the variety and growing stage of Perilla (Huang and Hwang, 1980). The pH differential method (Fuleki and Francis, 1968) was used to quantify anthocyanins. It was found that the anthocyanin content varied significantly with the variety of Perilla. Taiwan purple Perilla (*Perilla ocymoides* Linn. var. crispa Benth. forma purpurea Makino) contains 1.9g anthocyanins/100g of dried leaves, the highest among the three varieties under investigation. The change of the anthocyanin content with the growing stage was relatively small. *Perilla frutescens* Brit. var. crispa Decne, which was introduced from Japan, has smaller leaves than does Taiwan Perilla. It showed a remarkable change in the anthocyanin content with the growing stage. The anthocyanin content reached a maximum level (2.3g/100g) before flowering which was even higher than that of Taiwan purple Perilla. The average anthocyanin content, however, was 1.7g/100g. *Perilla frutescens* Brit. var. discolour Kudo was also introduced from Japan. It has even smaller leaves, and the leaves are reddish on one side only. As expected, this variety was found to contain the smallest amount of anthocyanins (1.2g/100g), and the change of the anthocyanin content with the growing stage was also the smallest.

The anthocyanin composition of Perilla has been studied extensively in Japan since 1931. A detailed review on the analysis of Perilla anthocyanins is presented in the previous chapter. Our laboratory has analysed the anthocyanins in the leaves of Taiwan purple Perilla (Wuu and Hwang, 1980). Eight anthocyanins were isolated and purified from Perilla leaves. They were identified by HPLC, TLC and paper chromatographic analyses of the aglycones, sugars and acyl groups, aided by spectral comparison and chromatographic co-elution with authentic compounds. The anthocyanins in Taiwan purple Perilla are, in order of decreasing abundance, cyanidin-3-(6-p-coumaryl-β-D-glucoside)-5-β-D-glucoside, cyanidin-3-(p-coumarylglucoside)-5-(p-coumarylglucoside), cyanidin-3, 5-diglucoside acylated with one molecule of coumaric acid, cyanidin-3, 5-diglucoside acylated with two molecules of coumaric acid, cyanidin-3-(caffeylglucoside)-5-(caffeylglucoside), cyanidin-3, 5-diglucoside acylated with one molecule of caffeic acid and two acylated cyanidin-3-glucosides.

EXTRACTION AND MEMBRANE CONCENTRATION OF PERILLA ANTHOCYANINS

Extraction

Anthocyanins are generally extracted with acidic alcohols. Methanol acidified with hydrochloric acid is the most effective extracting solvent (Francis, 1982). Acidified ethanol is slightly inferior, but it is preferred in food systems due to the toxicity of methanol. For the extraction of anthocyanins from raw plant materials, acidified water or water containing SO_2 (sulphur dioxide) is also used.

Buckmire and Francis (1978) extracted anthocyanins from miracle fruit with methanol containing 0.03% HCl. A similar solvent was also used in the extraction of cranberry anthocyanins (Chiriboga and Francis, 1973) and blueberry anthocyanins (Shewfelt and Ahmed, 1978). Extraction of cranberry anthocyanins was also achieved with 95% ethanol containing 1.5N HCl (Woo *et al.*, 1980). As for the extraction of grape anthocyanins, methanol containing 0.1% tartaric acid (Philip, 1974) or 0.01% citric acid (Clydesdale *et al.*, 1978) was used. In preparing grape anthocyanins for use as a food colorant, Main *et al.* (1978) employed ethanol containing 0.01% citric acid to extract grape waste. Besides acidified methanol and ethanol, hot water (Barel, 1978), acidified water (Clydesdale *et al.*, 1978) and water containing 500 ppm SO_2 (Palamidis and Markakis, 1975) were also

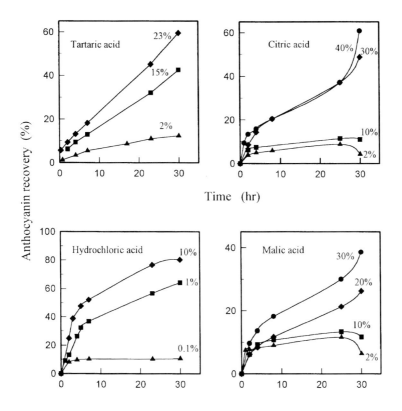

Figure 1   Extraction of anthocyanins from Perilla leaves with acidic ethanol solution

found effective in extracting anthocyanins from grape waste. Metivier *et al.*(1980) compared the extraction efficiency of methanol, ethanol and water containing different acids at various concentrations in the extraction of grape anthocyanins. Acidified methanol was found to give the highest yield while acidified water had the lowest.

Extraction of anthocyanins from Perilla leaves for analytical purposes has usually been achieved with acidified methanol (Hayashi and Abe, 1955; Wuu and Hwang, 1980) or acidic water as described in the previous chapter. In preparing Perilla anthocyanin extract for food use, Ryo (1974) patented an extracting solvent using 60° Brix aqueous sugar-citric acid solution at pH 1.6–3.0. Our laboratory has tried ethanol or water acidified with different concentrations of hydrochloric acid, citric acid, tartaric acid and malic acid. Dried Perilla leaves were chosen as the raw material since they have been shown to be the form which best preserves the colour of Perilla (Liu and Hwang, 1983). Dried leaves were immersed in solvent and stored at room temperature. Extract was analyzed periodically for the total anthocyanin extracted. Figure 1 shows that among the four acids, hydrochloric acid is most effective. It could recover more than 60% of the

LUCY SUN HWANG

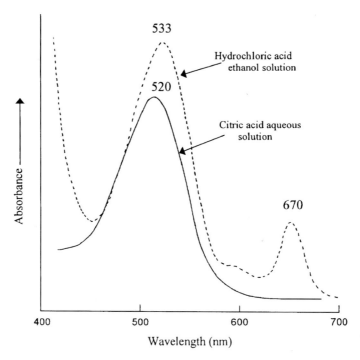

Figure 2 Absorption spectra of Perilla leaves extract in HCl ethanol solution and citric acid aqueous solution

anthocyanins, even at the 1% level, while a higher concentration is needed for other organic acids. Malic acid was found to be the least effective one; only 40% of the anthocyanins were recovered at the 30% malic acid level. Citric acid and tartaric acid were more effective than malic acid; nearly 60% of the anthocyanins could be recovered. A higher acid concentration was needed for citric acid (40%) than for tartaric acid (23%) to achieve this recovery level. When they extracted anthocyanins from wine pomace, Metivier *et al.* (1980) also had similar results.

Although acidic ethanol is quite effective in recovering anthocyanins from Perilla leaves, it is also capable of extracting chlorophylls. Being present in the leaves of Perilla, anthocyanins co-exist with chlorophylls. As shown in Figure 2, the HCl-ethanol extract of Perilla leaves has an absorption peak at 670 nm , corresponding to the chlorophyll absorption while the citric acid-water extract has no absorption in this region. Acidic water was, therefore, chosen to be the solvent system for extracting Perilla anthocyanins from dried Perilla leaves. Among the four different acids (mentioned above) used at various concentrations in water, hydrochloric acid was found to be least effective. The other three acids showed similar performances; all had better performance than in ethanol. For example, 30% of any acid could recover more than 80% of anthocyanins. Citric acid was selected in preparing Perilla anthocyanin extract for the later membrane concentration operation because of its low cost, good flavour and metal chelating ability.

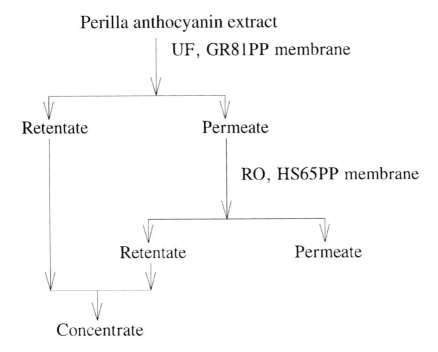

Figure 3 A schematic diagram of the concentration process of Perilla extract

Concentration

Anthocyanin extracts are usually concentrated by vacuum evaporation (Philip, 1974; Main *et al.,* 1978) or flash evaporation (Palamidis and Markakis, 1975). Perilla anthocyanins, however, are best extracted with high concentrations of a citric acid-water solution. Concentration by means of the membrane technique, which retains large molecules and allows small molecules to pass through the membrane, thus appeared to be a desirable choice. Perilla anthocyanin extract, prepared by extracting dried Perilla leaves with water containing 10% citric acid, was concentrated according to the scheme illustrated in Figure 3. The extract was first concentrated by ultrafiltration (UF), which could retain anthocyanins and allow citric acid to pass through, and the UF permeate was further concentrated by reverse osmosis (RO). In this study, an interesting phenomenon was observed. Perilla anthocyanins having molecular weights (MW) around 800 were retained by UF membrane DDS GR81PP which had MW cut-off = 6000. The retention of anthocyanin by this membrane was over 70%, and the retention of the total solids in the extract was only about 20%. This finding together with other evidence suggested that Perilla anthocyanins might have existed in the extract as copigmented complexes which were large enough to be retained by the UF membrane of MW cut-off 6000 (Chung *et al.,* 1986).

Recovery of Perilla anthocyanins was over 60% by UF at a volume concentration ratio (VCR=Vi/Vt, where Vi is the initial volume and Vt is the volume of concentrate) of 4. When the UF permeate was further concentrated by RO, even a highly permeable membrane (DDS HS65PP) was still not efficient in removing the solvent due to its high solid (citric acid) content. The UF permeate, which still contained some free anthocyanins and 9% citric acid was, therefore, possibly reused as an extracting solvent (Chung *et al.*, 1986).

In an attempt to improve the membrane concentration process, an alternative solvent system was used in our laboratory. After testing different concentrations of HCl (the smallest acid molecule) in water as an extracting solvent and investigating the best dried Perilla leaves (sample) to solvent ratio, an extraction system of 1% aqueous hydrochloric acid as solvent and a sample to solvent ratio of 50 g/L was established. The recovery of anthocyanins from dried Perilla leaves was around 75% after soaking for 8 hours, which was higher than that using 10% aqueous citric acid as solvent.

The 1% aqueous hydrochloric acid extract of Perilla anthocyanins was concentrated using the same membrane technology described in Figure 3. During the UF process, the retention of anthocyanins by DDS GR81PP membrane was above 87.5%. When the UF retentate was recycled through the UF membrane and concentrated continuously until VCR reached 10, the permeate flux (16.25 L/m^{2}hr) was still quite satisfactory (15 L/m^{2}hr is considered satisfactory in industrial operation).

As shown in Figure 4, the recovery of Perilla anthocyanins was nearly 85% at VCR = 4 and 76.6% at VCR = 10, which were much better than that using 10% aqueous citric acid as an extracting solvent. Upon concentrating the UF permeate by RO, the DDS HS65PP membrane was found to retain 96% of the Perilla anthocyanins. When the RO retentate was continuously recycled through the RO membrane until VCR reached 8, the recovery of anthocyanins was still above 90% (Figure 4). The final concentrate of Perilla anthocyanins after combining both the UF retentate and RO retentate (Figure 3) had an anthocyanin recovery of 88%. The anthocyanin concentration increased more than four-fold (from 218 mg/L in the original extract to 898 mg/L in the final concentrate) while the percentage of anthocyanin in the total solid also increased two-fold (from 0.83% to 1.59%).

From the results of the above extraction and concentration experiments, it is concluded that Perilla anthocyanins is more suitably extracted with 1% aqueous hydrochloric acid in order to achieve concentration by membrane processes at a satisfactory rate. Combining UF and RO processes in concentrating the Perilla anthocyanin extract could not only yield a concentrate with good pigment retention, but also could increase the purity of the pigment.

Recovery of Perilla Anthocyanins from Spent Brine by Membrane Processes

In Taiwan, Perilla leaves are generally preserved in salt and exported to Japan. The spent brine resulting from this pickling process, which contains around 230 mg/L anthocyanins, is a good source of Perilla anthocyanins. We have also investigated the possibility of recovering anthocyanins from this spent brine by means of diafiltration and electrodialysis (Lin, *et al.*, 1989).

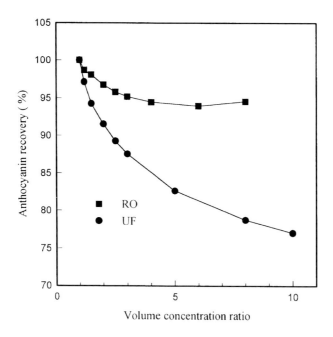

Figure 4   Anthocyanin recovery at different volume concentration ratios using GR81PP (UF) membrane at 11.5 bar pressure and HS65PP (RO) membrane at 29.35 bar pressure

Diafiltration is a modified UF process in which a diafiltering fluid (such as water) is added to the retentate to enhance the "wash out" of the membrane-permeable component in the solution (Beaton and Klinkowski, 1983). Since spent brine is very high in salt content (around 20%), this diafiltration technique may be more favourable than the regular UF process. The results of our study showed that 95% of the salt from the spent brine could be removed at an added-water volume to original feed ratio of 4. More than 90% of the Perilla anthocyanins could be recovered.

Another membrane process which may be used to remove salt from the spent brine is electrodialysis (ED). This process employs electrically-charged membranes to separate ions from a solution using an electrical potential difference as the driving force (Strathmann, 1985). Our study also demonstrated that the ED process could remove 90% of the salt while recovering 86% of the anthocyanins from the spent brine.

STABILITY OF PERILLA ANTHOCYANINS

Natural pigments are generally less stable than synthetic dyes. Being an important class of natural pigment, anthocyanins are no exception. The electron deficient flavylium nucleus on the anothocyanin molecule is highly reactive. Anthocyanins are, therefore, easily decolourised. The stability of anthocyanins has been extensively reviewed (Markakis, 1982; Jackman *et al.*, 1987; Mazza and Brouillard, 1987) and has received much attention

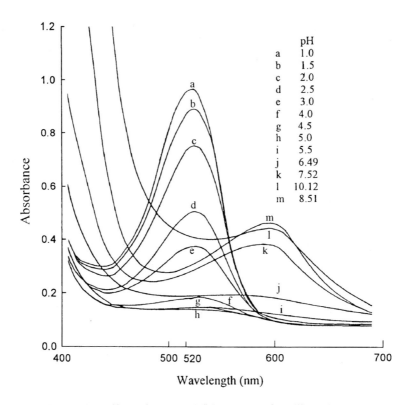

Figure 5   Effect of pH on visible spectra of Perilla anthocyanins

recently (Chandra *et al.*, 1993; Baublis, *et al.*, 1994; Cemeroglu *et al.*, 1994; Cheynier *et al.*, 1994; Barth *et al.*, 1995). The major factors affecting the degradation of anthocyanins are pH, oxygen, temperature, light, metals, polyphenols, ascorbic acid, and sugars and their degradation products. In order to understand the influences of the above factors on the stability of Perilla anthocyanins, the pigment concentrate prepared using the UF/RO process was diluted in buffer solution to 50 ppm (except in the copigmentation study) in the studies described below.

Effects of pH and Oxygen

Anthocyanins are very sensitive to pH changes. The pH value not only affects the colour of anthocyanins, but it also influences stability. The effect of pH on the absorption spectrum of Perilla anthocyanins is evident from Figure 5. The absorbance of visible absorption maximum at 520 nm decreased as the pH value increased, and a minimum absorbance was reached at around pH 5.0. A new absorption peak at 600 nm, which corresponded to the absorption of the quinoidal base of Perilla anthocyanin, started to emerge at pH values above 6.5.

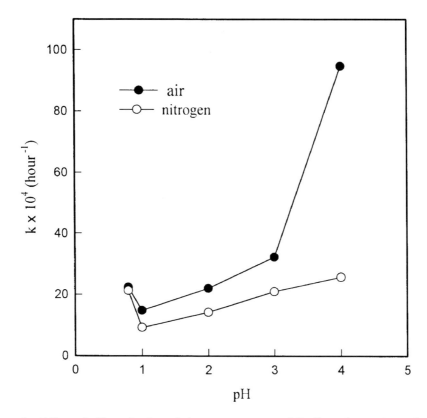

Figure 6   Effect of pH on the degradation rate constants of Perilla anthocyanins under air or nitrogen at 45°C

The effect of pH on the stability of Perilla anthocyanins was studied at 45°C under nitrogen in the dark. The degradation reaction was slowest at pH 1.0. When the pH was lower than 1.0, the degradation of anthocyanin was enhanced, possibly due to hydrolysis of the anthocyanin to less stable anthocyanidin (Maccarone *et al.*, 1985). The influence of pH on the anthocyanin stability was found to be oxygen dependent. Figure 6 shows the degradation reaction rates, calculated on the basis of first order reaction kinetics, at different pH values under air or nitrogen. It is clear that the influence of the pH value on the degradation of anthocyanins was less pronounced under nitrogen. Adams(1972) and Lukton *et al.* (1956) similarly observed that anthocyanin breakdown in the absence of air was virtually pH-independent in a pH range of 2 to 4. In this study, we also noticed that the degradation rate of anthocyanin at pH 0.8 was almost the same under either air or nitrogen.

It can, therefore, be concluded that oxygen plays an important role in the degradation of anthocyanin at pH values between 1 to 4. In order to improve the stability of Perilla anthocyanins at pH 2 and above, removal of oxygen may be beneficial.

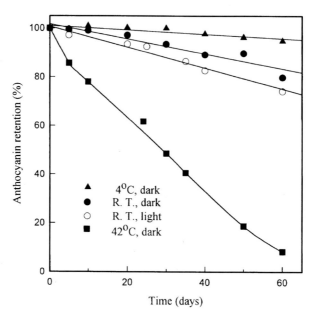

Figure 7   Effects of storage temperature and light on the stability of Perilla anthocyanins in pH 1.0 buffer

Effects of Temperature and Light

The degradation of anthocyanins during food processing and storage is markedly influenced by temperature either in natural or model systems (Daravingas and Cain, 1965; Hrazdina *et al.*, 1970; Shrikhande, 1976; Mishkin and Saguy, 1982; Maccarone *et al.*, 1985; Cemeroglu, *et al.*, 1994). The thermal stability of anthocyanins extracted from Perilla leaves with 1 % aqueous hydrochloric acid and concentrated by UF/RO was investigated in our laboratory after diluting the concentrate to 50 ppm.

The storage stability of Perilla anthocyanins was studied in pH 1.0 buffer, which is the most stable pH, in order to minimise the pH effect. Samples stored at 4°C in the dark were quite stable. At room temperature, Perilla anthocyanins could still retain 85 % pigment after storage for 2 months (Figure 7), the half life (time to destruct 50 % pigments) was around 200 days. When the storage temperature was raised to 42°C, the degradation of anthocyanin was enhanced intensely, the half life was only 25 days. Iacobucci and Sweeny (1983) reported that the half life of cyanidin-3-rutinoside was about 65 days at room temperature. The half life of strawberry anthocyanins was 10 days at 38°C and 54 days at 20°C (Meschter, 1953). Compared to these anthocyanins, Perilla anthocyanins are relatively stable.

Stability of Perilla anthocyanins at temperatures usually encountered during food processing was also studied. It was investigated in pH 3.0 buffer solution and in a simulated soft drink containing 13% sucrose, 0.1% citric acid, 0.05% ascorbic acid, 0.06% sodium benzoate and buffered at pH 3.0. The degradation reaction rate constants and half life values, calculated on the basis of first order reaction kinetics, at temperatures

Table 1 Degradation rate constants (k), half - life values $(T_{1/2})$ and activation energies (E_a) of perilla anthocyanins in pH 3.0 buffer solution and a simulated soft drink at 60, 80, 100, and 121°C

Temperature	*Anthocyanin in pH 3.0 buffer solution*		*Anthocyanin in pH 3.0 simulated drink*	
°C	k (min^{-1})	$T_{1/2}$ (min)	k (min^{-1})	$T_{1/2}$ (min)
60	0.000297	2336	0.001111	624
80	0.001596	434	0.002234	310
100	0.003905	177	0.004317	161
121	0.030013	23	0.031024	22
Ea(kcal / mole)	19.01		13.40	

from 60 to 121°C are listed in Table 1 together with the activation energies. In the simulated soft drink system, Perilla anthocyanins had faster degradation rate and lower activation energy than in pure buffer system due to the presence of ascorbic acid and sucrose, which are known to destroy anthocyanins (Daravingas and Cain, 1965; Jackman *et al.*, 1987). The activation energy of grape anthocyanins in the soft drink model system was reported to be 13.74 kcal / mole (Palamidis and Markakis, 1975). Although Perilla anthocyanins had a similar activation energy (13.40 kcal / mole), it contained ascorbic acid in the simulated soft drink while the grape anthocyanin system did not have ascorbic acid. Therefore, Perilla anthocyanins are possibly more stable toward food processing than grape anthocyanins.

The effect of light on anthocyanins are two directions. It favours the biosynthesis of anthocyanins, but it will also accelerate their degradation (Markakis, 1982). The chemical structure of anthocyanin was reported to influence the adverse effect of light on anthocyanins (Van Buren *et al.*, 1968; Attoe and von Elbe, 1981; Sapers, *et al.*, 1981). As is evident from Figure 7, light had a rather limited effect on the storage stability of Perilla anthocyanins at room temperature. The calculated half life was 146 days under light and 200 days in the dark. Light only accelerated 1.3 times the destruction of Perilla anthocyanins, while it speeded up the decomposition of grape anthocyanins by 2 times (Palamidis and Markakis, 1975). Van Buren *et al.* (1968) reported that acylated, methylated diglycosides were the most light stable anthocyanins, followed by nonacylated diglycosides and the monoglycosides were the least stable ones. Being mostly acylated diglycosides, Perilla anthocyanins are thus more stable to light.

Effects of Polyphenols and Metals

Anthocyanins are well known to form weak complexes with polyphenolic compounds, such as flavonoids and gallotannins, polysaccharides, metals and other substances, a phenomenon known as copigmentation (Asen *et al.*, 1972; Markakis, 1974; Brouillard *et al.*, 1989; Mazza and Brouillard, 1990; Goto and Kondo, 1991; Davies and Mazza, 1993). This phenomenon plays a major role in the expression of the wide range of brilliant colours displayed by anthocyanins. Copigmentation results in a bathochromic shift in the visible λ_{max} (wavelength of maximum absorption), from red to blue, as well as an increase in the absorbance at λ_{max} (Scheffeldt and Hrazdina, 1978; Osawa, 1982; Brouillard et al., 1989). Anthocyanins existing in copigmentation forms may also have greater stabilities (Shrikhande and Francis, 1974; Maccarone et al., 1985).

The possibility of copigmentation between Perilla anthocyanins and flavonoids was investigated. Rutin, one of the most efficient copigments (Somers and Evans, 1977), was added to Perilla anthocyanins (acy) at two molar ratios (rutin / acy = 6 and 24) in ten different buffer solutions from pH 1.0 to 5.5. There was no shift in λ_{max} nor increase in absorbance at λ_{max}, that is, no copigmentation was observed between Perilla anthocyanins and rutin. Caffeic acid, which was found to display copigmentation with acylated pelargonidin (Davies and Mazza, 1993), was also tested under the same conditions as rutin. Again, there was no sign of copigmentation.

Anthocyanins may also be stabilised by forming metal complexes with iron, tin, copper and aluminium (Markakis, 1974). In general, metals are not necessary for copigmentation to proceed (Asen et al., 1972). Jurd and Asen (1966), however, reported that Al^{+3} might help the copigmentation between cyanidin-3-glucoside and quercitrin by acting as cross-linking agent. Perilla anthocyanins were, therefore, mixed with $AlCl_3$ and rutin in different buffer solutions. As is clear from Table 2, the addition of $AlCl_3$ in the buffer solution containing only anthocyanin did show a bathochromic shift of λ_{max} and an increase in absorbance at λ_{max} in pH 2.5 and pH 3.0. This was caused by metal complex formation between Al^{+3} and Perilla anthocyanins which had o-diphenolic group in the β ring of anthocyanin molecule. No additional shift in λ_{max} was observed when rutin was added. Therefore, $AlCl_3$ did not exert any effect on copigmentation formation between Perilla anthocyanins and rutin.

From the above experiments, it is concluded that Perilla anthocyanins can form complex with Al^{+3}, but they cannot form copigmentation with flavonoids even with the aid of Al^{+3}. It is postulated that Perilla anthocyanins have already existed as intramolecular copigmentation or intermolecular copigmentation with naturally present flavonoids under the experimental condition (anthocyanin concentration = 4.5×10^{-4} M), therefore, additional flavonoids cannot induce more copigmentation. The shift of λ_{max} from 531 nm to 541 nm when the pH value of Perilla anthocyanins increased from pH 1.0 to 3.0 (Table 2) seemed to support this postulation.

Table 2 Effects of rutin and $AlCl_3$ on the color of perilla anthocyanins at different pH values

pH	Acy in buffer		Acy in buffer + Rutin*		Acy in buffer +AlCl₃**		Acy in buffer +AlCl₃** + Rutin*	
	λ_{max}	$A_{\lambda max}$	λ_{max}	$A_{\lambda max}$	λ_{max}	$A_{\lambda max}$	λ_{max}	$A_{\lambda max}$
1.0	531	1.407	533	1.275	530	1.316	533	1.361
1.5	532	1.388	534	1.418	531	1.393	533	1.342
2.0	534	1.313	536	1.322	532	1.307	535	1.301
2.5	538	1.266	538	1.274	552	1.274	552	1.295
3.0	541	1.167	540	1.184	568	1.435	568	1.423

* : Rutin / Acy (molar ratio) = 24
** : AlCl₃ / Acy (molar ratio) = 1000

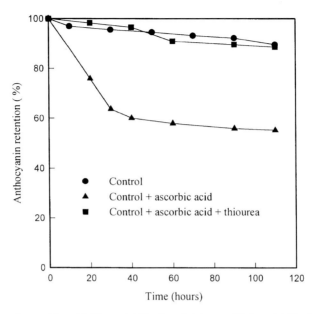

Figure 8   Effects of ascorbic acid (50mg / 100ml) and thiourea (100mg / 100ml) on the pigment retention of Perilla anthocyanins (50ppm) in pH 3.0 buffer solution at 30°C in the presence of air.

Effect of Ascorbic acid

Natural fruit juices contain ascorbic acid. Ascorbic acid may also be used to fortify beverages. Beattie *et al.* (1943) were among the first researchers to observe the concurrent disappearance of ascorbic acid and anthocyanins in stored fruit juices. Sondheimer and Kertesz (1953) suggested that maximal loss of anthocyanins occurred under conditions most favourable to ascorbic acid oxidation. Calvi and Francis (1978) found that oxygen would accelerate the destruction of anthocyanins by ascorbic acid.

Inevitably, Perilla anthocyanins are also susceptible to destruction by ascorbic acid. We have tried to retard this reaction by incorporating thiourea, which is known to protect ascorbic acid from oxidation (Skalski and Sistrunk, 1973), in anthocyanin solution containing ascorbic acid. The results are shown in Figure 8. Thiourea added at 100mg/ 100ml level to Perilla anthocyanins solution (50 ppm in pH 3.0 buffer) containing ascorbic acid (50 mg/100ml) indeed retained as much anthocyanin as the control group which contained no ascorbic acid.

The protective effect of thiourea was further investigated under both aerobic and anaerobic conditions at 60°C in pH 3.0 buffer. It is evident from Figure 9 that under aerobic conditions(curves with white marks), the destruction of anthocyanins was more severe and the protection of thiourea was more efficient than anaerobic conditions. The reaction between ascorbic acid and Perilla anthocyanins under anaerobic conditions in pH 1.0, 3.0 and 7.0 buffer solutions was also conducted in our laboratory. It was found that the destruction of anthocyanins was most severe in pH 1.0 solution followed

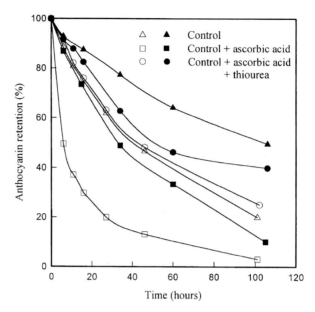

Figure 9   Effects of ascorbic acid (50mg / 100ml) and thiourea (100mg / 100ml) on the pigment retention of Perilla anthocyanins (50 ppm) in pH 3.0 buffer solution at 60°C in the presence or absence of air. (White marks: aerobic, Black marks: anaerobic)

by pH 3.0, and pH 7.0 solution had the least destruction of anthocyanins. This was quite in contrary to the stability of anthocyanins in different buffer solutions (Figure 6). It was, however, in accordance with the degradation of ascorbic acid under anaerobic conditions (Tannenbaum and Young, 1985). The results of the above studies strongly suggested that Perilla anthocyanins were destroyed by the degradation products of ascorbic acid instead of reacting with ascorbic acid directly.

CONCLUSION

Perilla leaf is a rich source of anthocyanins. The common solvents used for extracting anthocyanins, acidic alcohols, were found to extract chlorophylls simultaneously. Perilla anthocyanins are, therefore, more suitable to be extracted from dried leaves with aqueous citric acid or aqueous hydrochloric acid. Aqueous hydrochloric acid is the preferred extracting solvent since it can facilitate the membrane concentration process. Concentration of the Perilla anthocyanin extract with UF / RO process can also increase the purity of anthocyanin since UF membrane can retain more anthocyanins than the smaller molecules in the extract. Perilla anthocyanins may possibly exist in the extract as copigmentation complex and thus can be retained by UF membrane. Perilla anthocyanins can also be recovered from the spent brine of pickled Perilla by diafiltration or electrodialysis, thus providing a cheap source of Perilla anthocyanins.

Similar to most anthocyanins, Perilla anthocyanins are sensitive to pH, oxygen, heat and ascorbic acid. Removal of oxygen may improve the stability of Perilla anthocyanins at pH 2 and above. Perilla anthocyanins were found to be more stable during thermal processing than grape anthocyanins. Perilla anthocyanins are also more stable toward light since they are acylated anthocyanins. Destruction of Perilla anthocyanins by ascorbic acid was more severe under aerobic condition than anaerobic condition. Thiourea was able to inhibit this destruction reaction. The degradation products of ascorbic acid was suggested to be the active principles in destroying anthocyanins. Perilla anthocyanins can form metal complexes with $AlCl_3$. Copigmentation between Perilla anthocyanins and rutin was not observed, even with the aid of Al^{+3}. This again indicated that Perilla anthocyanins might already form intra- or inter- molecular copigmentation complexes with naturally occurring flavonoids.

REFERENCES

Adams, J.B. (1972) Changes in the polyphenols of red fruits during processing – the kinetics and mechanism of anthocyanin degradation. *Campden Food Preserv. Res. Assoc. Tech. Bull.*, pp. 22.

Asen, S., Stewart, R.N. and Norris, K.H. (1972) Copigmentation of anthocyanins in plant tissues and its effect on color. *Phytochem.*, **11**, 1139–1144.

Attoe, E.L. and von Elbe, J.H. (1981) Photochemical degradation of betanine and selected anthocyanins. *J. Food Sci.*, **46**, 1934–1937.

Barel, H.S. (1978) Extraction of anthocyanins from marc of grapes, lees and wines. *Fr. Demande*, **2**, 378070.(CA **90**, 85286d).

Barth, M.M., Zhou, C., Mercier, J., and Payne, F.A. (1995) Ozone storage effects on anthocyanin content and fungal growth in blackberries. *J. Food Sci.*, **60**, 1286–1288.

Baublis, A., Spomer, A. and Berber-Jimenez, M.D. (1994) Anthocyanin pigments: Comparison of extract stability. *J. Food Sci.*, **59**, 1219–1221.

Beaton, N.Y. and Klinkowski, P.R. (1983) Industrial ultrafiltration design and application of diafiltration processes. *J. Separ. Proc. Technol.*, **4**, 1–10.

Beattie, H.G., Wheeler, K.A. and Pederson, C.S. (1943) Changes occurring in fruit juices during storage. *Food Res.*, **8**, 395–404.

Brouillard, R., Mazza, G., Saad, Z., Albrecht-Gary, A.M., Cheminat, A. (1989) The co-pigmentation reaction of anthocyanins: A microprobe for the structural study of aqueous solutions. *J. Am. Chem. Soc.*, **111**, 2604–2610.

Buckmire, R.E. and Francis, F. J. (1978) Pigments of miracle fruit, Synsepalum dulcificum, Schum, as potential food colorants. *J. Food Sci.*, **43**, 908–911.

Calvi, J.P. and Francis, F.J. (1978) Stability of concord grape (V. labrusca) anthocyanins in model systems. *J. Food Sci.*, **43**, 1448–1456.

Cemeroglu, B., Velioglu, S. and Isik, S. (1994) Degradation kinetics of anthocyanins in sour cherry juice and concentrate. *J. of Food Sci.*, **59**, 1216–1218.

Chandra, A., Nair, M.G. and Iezzoni, A.F. (1993) Isolation and stabilization of anthocyanins from tart cherries (Prunus cerasus L.). *J. Agric. Food Chem.*, **41**, 1062–1065.

Cheynier, V., Souquet, J., Kontek, A., Moutounet, M. (1994) Anthocyanin degradation in oxidising grape musts. *J. Sci. Food Agric.*, **66**, 283–288.

Chiriboga, C.D. and Francis, F.J. (1973) Ion exchange purified anthocyanin pigments as a colorant for cranberry juice cocktail. *J. Food Sci.*, **38**, 464–467.

Chung, M.Y. , Hwang, L.S. and Chiang, B.H. (1986) Concentration of perilla anthocyanins by ultrafiltration. *J. Food Sci.*, **51**, 1494–1497, 1510.

Clydesdale, F.M., Main, J.H., Francis, F.J., Damon, Jr., R.A. (1978) Concord grape pigments as colorants for beverages and gelatin desserts. *J. Food Sci.*, **43**, 1687–1692.

Daravingas, G. and Cain, R.F. (1965) Changes in the anthocyanin pigments of raspberries during processing and storage. *J. Food Sci.*, **30**, 400–405.

Davies, A.J. and Mazza, G. (1993) Copigmentation of simple and acylated anthocyanins with colorless phenolic compounds. *J. Agric. Food Chem.*, **41**, 716–720.

Francis, F.J. (1982) Analysis of anthocyanins. In P. Markakis (ed.), *Anthocyanins as food colors*, Academic Press, New York, pp. 181–207.

Fuleki, T. and Francis, F.J. (1968) Quantitative methods for anthocyanins. 2. Determination of total anthocyanin and degradation index for cranberry juice. *J. Food Sci.,* **33**, 78–83.

Goto, T. and Kondo, T. (1991) Structure and molecular stacking of anthocyanins-flower color variation. *Angew. Chem., Int. Ed. Engl.*, **30**, 17–33.

Hayashi, K. and Abe, Y. (1955) Studies on anthocyanins XXV: paper chromatographic investigation on anthocyanins occurring in the leaves of Perilla varieties. *Bot. Mag. Tokyo*, **68**, 71–75.

Hrazdina, G., Borzell, A. J. and Robinson, W. B. (1970) Studies on the stability of the anthocyanidin-3, 5-diglucosides. *Am. J. Enol. Vitic.*, **21**, 201–204.

Huang, S.A. and Hwang, L.S. (1980) Studies on the natural red pigment of perilla: changes of anthocyanin content with variety and growing stage. *Food Science* (Taiwan), **7**, 161–169.

Iacobucci, G.A. and Sweeny, J.G. (1983) The chemistry of anthocyanins, anthocyanidins and related flavylium salts. *Tetrahedron*, **39**, 3005–3008.

Jackman, R.L., Yada, R.Y., Tung, M.A., Speers, R.A. (1987) Anthocyanins as food colorants-a review. *J. Food Biochem.*, **11**, 201–247.

Jurd, L. and Asen, S. (1966) The formation of metal and "copigment" complexes of cyanidin 3-glucoside. *Phytochem.*, **5**, 1263–1271.

Liu, C.H. and Hwang, L.S. (1983) Studies on the natural red pigment of perilla: changes of anthocyanin content during some post-harvest treatments. *Food Science* (Taiwan), **10**, 43–53.

Lin, S.S., Chiang, B.H. and Hwang, L.S. (1989) Recovery of perilla anthocyanins from spent brine by diafiltration and electrodialysis. *J. Food Engineering*, **9**, 21–33.

Lukton, A., Chichester, C.O. and Mackinney, G. (1956) The breakdown of strawberry anthocyanin pigment. *Food Tech.*, **10**, 427–432.

Maccarone, E., Maccarrone, A. and Rapisarda, P. (1985) Stabilization of anthocyanins of blood orange fruit juice. *J. Food Sci.*, **50**, 901–904.

Main, J.H., Clydesdale, F.M. and Francis, F.J. (1978) Spray drying anthocyanin concentrates for use as food colorants. *J. Food Sci.*, **43**, 1693–1694.

Markakis, P. (1974) Anthocyanins and their stability in foods. *CRC Crit. Rev. in Food Technol.*, **4**, 437–456.

Markakis, P. (1982) Stability of anthocyanins in foods. In P. Markakis, (ed.), *Anthocyanins as Food Colors*, Academic Press, New York, pp. 163–180.

Mazza, G. and Brouillard, R. (1987) Recent developments in the stabilization of anthocyanins in food products. *Food Chem.*, **25**, 207–225.

Mazza, G. and Brouillard, R. (1990) The mechanism of co-pigmentation of anthocyanins in aqueous solutions. *Phytochem.*, **29**, 1097–1102.

Meschter, E.L. (1953) Fruit color loss: Effect of carbohydrate and other factors on strawberry products. *J. Agric. Food Chem.*, **1**, 574–579.

Metivier, R.P., Francis, F.J. and Clydesdale, F.M. (1980) Solvent extraction of anthocyanins from wine pomace. *J. Food Sci.*, **45**, 1099–1100.

Mishkin, M. and Saguy, I. (1982) Thermal stability of pomegranate juice. *Z. Lebensm. u. Forsch.*, **175**, 410–412.

Osawa, Y. (1982) Copigmentation of anthocyanins. In P. Markakis, (ed.), *Anthocyanins as Food Colors*, Academic Press, New York, pp. 41–68.

Palamidis, N. and Markakis, P. (1975) Stability of grape anthocyanin in a carbonated beverage. *J. Food Sci.*, **40**, 1047–1049.

Philip, T. (1974) An anthocyanin recovery system from grape wastes. *J. Food Sci.*, **39**, 859.

Ryo, A. (1974) Food additive from perilla. *Japan Kokai,* **74**, 86, 516. (CA **85**, 4082f).

Sapers, G.M., Taffer, I. and Ross, L.R. (1981) Functional properties of a food colorant prepared from red cabbage. *J. Food Sci.*, **46**, 105–109.

Scheffeldt, P. and Hrazdina, G. (1978) Copigmentation of anthocyanins under physiological conditions. *J. Food Sci.*, **43**, 517–520.

Shewfelt, R.L. and Ahmed, E.M. (1978) Enhancement of powdered soft drink mixes with anthocyanin extracts. *J. Food Sci.*, **43**, 435–438.

Shrikhande, A.J. (1976) Anthocyanins in foods. *CRC Crit. Rev. in Food Sci. and Nutrition*, **7**, 193–218.

Shrikhande, A.J. and Francis, F.J. (1974) Effect of flavonols on ascorbic acid and anthocyanin stability in model systems. *J. Food Sci.*, **39**, 904-906.

Skalski, C. and Sistrunk, W.A. (1973) Factors influencing color degradation in concord grape juice. *J. Food Sci.*, **38**, 1060–1062.

Somers, T.C. and Evans, M.E. (1977) Spectral evaluation of young wines: Anthocyanin equilibria, total phenolics, free and molecular sulfur dioxide "chemical age". *J. Sci. Food Agric.*, **28**, 279–287.

Sondheimer, E. and Kertesz, Z.I. (1953) Participation of ascorbic acid in the destruction of anthocyanin in strawberry juice and model systems. *Food Res.*, **18**, 475–479.

Strathmann, H. (1985) Electrodialysis and its application in the chemical process industry. *Sep. Purification Methods*, **14**, 41–66.

Tannenbaum, S.R. and Young, V.R. (1985) Vitamins and Minerals. In O. R. Fennema, (ed.), *Food Chemistry*, Marcel Dekker, Inc., New York, pp. 478–544.

Van Buren, J.P., Bertino, J.J. and Robinson, W.B. (1968) The stability of wine anthocyanins on exposure to heat and light. *Am. J. Enol. Vitic.*, **19**, 147–154.

Woo, A.H., von Elbe, J.H. and Amundson, C.H. (1980) Anthocyanin recovery from cranberry pulp wastes by membrane technology. *J. Food Sci.*, **45**, 875–879.

Wuu, C.F. and Hwang, L.S. (1980) Anthocyanins in perilla leaves. *Food Science* (Taiwan), **7**, 60–82.

INDEX